Baas Becking's: Geobiology

Baas Becking's: Geobiology

Or Introduction to Environmental Science

Dr. L. G. M. Baas Becking (1895–1963)
Professor at the University of Leiden

EDITED BY

Don E. Canfield

TRANSLATED BY

Deborah Sherwood
Mishka Stuip

WILEY Blackwell

Registered Office
John Wiley & Sons, Ltd, The Atrium, Southern Gate, Chichester, West Sussex, PO19 8SQ, UK

Editorial Offices
9600 Garsington Road, Oxford, OX4 2DQ, UK
The Atrium, Southern Gate, Chichester, West Sussex, PO19 8SQ, UK
111 River Street, Hoboken, NJ 07030-5774, USA

For details of our global editorial offices, for customer services and for information about how
to apply for permission to reuse the copyright material in this book please see our website at
www.wiley.com/wiley-blackwell.

Translated by Deborah Sherwood and Mishka Stuip

Library of Congress Cataloging-in-Publication Data

Baas Becking, L. G. M., 1895–1963.
 [Geobiologie of inleiding tot de milieukunde. English]
 Baas Becking's Geobiology / Dr. L.G.M. Baas Becking, professor at the University of Leiden ;
edited by Don E. Canfield.
   pages cm
 Originally published in Dutch as: Geobiologie of inleiding tot de milieukunde (The Hague, the
Netherlands : W.P. Van Stockum & Zoon, 1934).
 Includes bibliographical references and index.
 ISBN 978-0-470-67381-2 (cloth)
1. Geobiology. 2. Aquatic ecology. 3. Marine biology. 4. Marine chemical ecology.
I. Canfield, Donald E., editor of compilation. II. Title. III. Title: Geobiology.
 QH343.4.B3313 2015
 577.6–dc23

                                        2015019931

A catalogue record for this book is available from the British Library.

Wiley also publishes its books in a variety of electronic formats. Some content that appears in print
may not be available in electronic books.

Cover image: A gypsum-encrusted microbial phototrophic community from the evaporating ponds
of a salt-production facility in Eilat, Israel © Don E. Canfield

Set in 9/11pt Meridien by SPi Global, Pondicherry, India

1   2016

Contents

Editor's Introduction

Lourens G. M. Baas Becking was born on January 4, 1895, in the town of Deventer, the Netherlands.[1] Little is written of Baas Becking's childhood, but early in his career he studied at the Technical University of Delft under the influence of Martinus W. Beyerinck. These were heady times at Delft. In the late 1800s Beyerinck was instrumental in establishing virology as field, and, perhaps of even greater importance, he perfected and applied the enrichment culture technique to understanding the diversity of microbes and microbial metabolisms in nature. Along the way, he discovered both nitrogen fixation and sulfate reduction. Baas Becking, however, did not finish his studies at Delft. Instead, he switched to the University of Utrecht, where he studied botany under the tutelage of Friedrich August Ferdinand Christian Went. His PhD studies included a visit to Stanford, and on his return to Utrecht he delivered his thesis, "Radiation and Vital Phenomena," published as a well-known book that is easily available today from multiple booksellers. After delivering his thesis, Baas Becking returned to Stanford as an assistant professor.

Baas Becking's time in California was of huge importance to his development of the concept of geobiology. He became especially fascinated with saline lakes and the adaptation of organisms to extreme environments. He was a keen observer of nature, unraveling how elements became biologically cycled and how this cycling both responded to and influenced the chemical environment. One can link this approach directly with Baas Becking's early experience at Delft and his exposure to Beyerinck's view of how chemical and physical factors (such as temperature) controlled the distribution of organisms in nature. The coupling of these insights, led to the famous quote *"everything is everywhere*: but *the environment selects,"* as developed in Chapter II of this book. Indeed, this sentiment in general, and the interplay between organisms and the chemical environment in particular, provided the foundation for Baas Backing's view of geobiology. Overall, Baas Becking's California experience is well expressed in the pages of this book.

While at Stanford, Baas Becking was appointed director of the Jacques Loeb Marine Laboratory in Pacific Grove, California, but he elected in 1930 to return to Leiden as professor of general botany. He continued his work on the

---

[1]In summarizing the history of Baas Becking's life, I have borrowed heavily from: Quispel, A. (1998) Lourens G. M. Baas Becking (1895–1963), Inspirator for many (micro) biologists, *International Microbiology*, 1, 69–72, and Oren, A. (2011) The halophilic world of Lourens Baas Becking. In Ventosa, A. *et al.* eds, *Halophiles and Hypersaline Environments*, Springer-Verlag, Berlin, pp. 9–26.

adaptations of organisms to the environment and summarized his thinking in a series of lectures to the scientific society "Diligentia" in The Hague, and these lectures formed the basis of the present book, *Geobiologie of inleiding tot de milieukunde*, published in 1934.

After returning to Leiden, Baas Becking continued his work on salt environments, on photosynthesis, and on symbiotic associations. His zest for botany, however, remained, and while researching geobiological themes he also oversaw the restoration of the botanical gardens in Leiden. Continuing on this theme, he moved with his family to Java in 1940 to help restore the Botanical Garden of Buitenzorg. He returned to Leiden to deliver his valedictory lecture on April 24, 1940, just before the Nazi invasion of the Netherlands. Unable to travel, he was separated from his family for the next five years. During the occupation he tried twice to escape, was imprisoned, and was nearly executed. In regaining contact with his family after the war, he learned of the tragic death of his son in battle on the Java Sea. Tragedy continued in 1948 with the death of his wife in an auto accident, shortly after he had moved to New Caledonia to take up a post with the newly established South Pacific Commission.

From New Caledonia, Baas Becking moved to Australia, occupying a number of positions before passing away on January 6, 1963. In recognition of his great contribution to Australian science, and to geobiology as a field, the Bureau of Natural Resources opened in 1965 the Baas Becking Geobiological Laboratory. This was, as far as I know, the first dedicated geobiological laboratory in the world. The initial focus of the Baas Becking labs was to explore how biological and chemical processes conspired to form stratiform ore deposits. By the time I learned of research from the Baas Becking Labs, the mission had broadened. When I was doing my PhD, the laboratory was quite famous for its work on modern stromatolites, and how these informed the interpretation of their ancient counterparts.

Sadly, the Baas Becking labs closed down in 1988. Ironically, this was some 10–15 years before geobiology began to emerge as a recognized scientific discipline with dedicated university programs, its own journal, and high exposure in international meetings such as the annual Goldschmidt Conference. Laurens Baas Becking and his legacy institute were well ahead of their time.

The current volume, or in reality collection of essays, elaborates on the geobiology of a wide variety of natural systems. In coining the term *geobiology*, Baas Becking was careful to insist that he was not trying to invent a new field, but rather, he was merely trying to express the relationship between organisms and the Earth. Perhaps this modesty respected earlier contributions of those like Vladimir Vernadsky, and particularly Sergei Vinogradsky, who, each in their own way, shared similar views to Baas Becking. Or perhaps this modesty was part of his character. In any event, these pages reveal a carefully developed consideration of the "geobiology" of organisms. The focus in these pages is largely, but not exclusively, on microorganisms, and one can clearly feel the influence of Beyerinck throughout the essays. Also, in a nod to Vinogradsky, and perhaps a nod away from Vernadsky, Baas Becking is almost exclusively concerned with modern environments. Therefore, the "geo" in Baas Becking's "geobiology" is the geosphere as it reflects a collection of environments, but it only casually considers (in Chapter VI) geologic time and geologic processes. Indeed, Baas Becking's "geobiology" has much in common with modern environmental microbial ecology and could equally be viewed as an instigator of this important field.

The book starts with a general introduction (Chapter I), a rather philosophical discourse on nature and science that ends with Baas Becking's definition of geobiology. In Chapter II, on the environment, he explores the metabolic diversity of life, as it was known then, and develops, with heavy inspiration from Beyerinck, the concept of *"everything is everywhere..."* as discussed above. Light is the focus of Chapter III, where Baas Becking explores how the availability of light impacts photosynthetic organisms in nature, a subject of particular interest to him given his background as a botanist. Indeed, in this chapter, Baas Becking makes some remarkable conjectures as to how photosynthetic organisms, like anoxygenic phototrophs for example, are well adapted to the spectra of light available to them. Chapter IV discusses temperature as it influences the physical environment, but of most interest (to me anyway) is Baas Becking's exploration of how organisms are adapted to the temperature of their environment. This chapter clearly demonstrates the author's interest in how organisms adapt to extreme environments. Chapter V then examines the chemical environment, with topics ranging from gases, to pH and how it is controlled, to the limiting constituents of life. What is inspiring here is Baas Becking's continued exploration of how organisms both respond to and control the chemical environment. Chapter VI considers element "cycles," and provides a remarkable integration of available knowledge (some of it just emerging at the time) to draw now iconic cartoon visualizations of the carbon, nitrogen, and sulfur cycles. It is both humbling and inspiring to appreciate how much Baas Becking knew about the biological control of element cycling.

These first chapters, in sum, could be viewed as the book's introduction. The following chapters explore diverse environments, including oligotrophic waters (Chapter VII), eutrophic fresh water (Chapter VIII), the oceans (Chapter IX), and brine (Chapter X). Each of these offers fascinating insights into the relationship between organisms and the environment, with, in many cases, observations from his own work and the work of his immediate colleagues. It is clear that, of these, "brine" was Baas Becking's favorite topic, with this chapter by far the longest and most comprehensive in the book.

Overall, Baas Becking provides us with a wonderful series of essays clearly establishing geobiology as a new integrated way to view organisms in their various natural environments. It is a pity that these essays, with their key insights, have been so long unavailable to other than the Dutch-speaking world. This, however, has changed with this new translated volume. Therefore, I wish first to thank Deborah Sherwood and Mishka Stuip for their expert translation of Baas Becking's *Geobiologie* into English. These two translators took a strong personal interest in producing the best and most accurate translation possible. They have also tried, as far as they could, to preserve Baas

Becking's original language, which was quite "old fashioned" and sometimes difficult to follow. My job was to check the translations for meaning, and the language for "flow" (within reason, given the primary objective to preserve Baas Becking's voice). I have also added some footnotes offering expansions and clarifications of Baas Becking's text, and each chapter is finished with "editor's notes" where I try to summarize the high points of the chapter and to put them into a modern context. Ultimately, any inconsistencies in style in the text and inaccuracies in meaning are my responsibility alone.

I would further like to thank Peter Westbrook for finding these two talented translators and for his great encouragement of this project. I wish to thank the Baas Becking family for permission to produce this translation and for general encouragement. I also wish to thank Ian Francis of Wiley for his encouragement and patience and the drafts people at Wiley-Blackwell for their excellent reproductions of Baas Becking's original figures. The translations were financed by the Royal Netherlands Academy of Arts and Sciences and the Agouron Institute, where I would particularly like to thank John Abelson for his enthusiasm for this project and for his general service to geobiology as a field. I would also like to thank the Danish National Research Foundation and the European Research Council (ERC)for support, and Aharon Oren for help with historical facts regarding Baas Becking's life. Finally, it has been a privilege to be involved in this project and to be the first non-Dutch-speaker to read these priceless words.

Don E. Canfield
Odense, Denmark

CHAPTER I

Introduction

The natural environment is something completely different to the observant enthusiast – Jules Renard's *chasseur d'images*[1] – than it is to the researcher attempting to discover the order and patterns of various life forms in a labora- tory. The former, the field biologist, views nature as an enormous, composite work of art. This manner of regarding nature is wholly satisfactory, so long as one is concerned only with "what." One can name the organisms he sees and, led in part by intuition, one can likewise appreciate systems of organisms: communities of life forms that are often associated with a certain type of land- scape (heath, dune, etc.). One can distinguish these communities even further and note the propagation of certain organisms (for example, the spread of cross-leaved heath in a field of heather).

However, in addition to the question of "what," human beings also ask the questions of "how" and "why" (53),[2] because the "what" – necessary as it may be – leads only to cataloging inventories, and by simply naming the parts one can never learn to understand the whole.

One can attempt to answer the question of "how," the second step toward understanding the natural world around us, without experimental tools. Yet in doing so, one encounters great difficulties. Both the external surroundings and the internal properties of organisms, which make possible the existence of large natural systems, often resist even the simplest attempts to analyze them. In order to study organisms, one must first study their environment. This can only be done in a place where the environment can be controlled, namely in a laboratory. Thus, under certain circumstances, an analysis is made of the relation between certain organisms and the controlled laboratory environment. Such an environment can be *homogeneous*, i.e., the external conditions in the experimental space either remain constant or change continually. When carrying out such an analysis in the field, one stumbles upon larger, in most

---

[1] Jules Renard (1864–1910) was a highly influential French author and keen observer of the natural world.
[2] Numbers placed between parentheses refer to the References at the end of this book.

Baas Becking's: Geobiology, Or Introduction to Environmental Science, First Edition. Edited by Don E. Canfield.
© 2016 John Wiley & Sons, Ltd. Published 2016 by John Wiley & Sons, Ltd.

cases even insurmountable, difficulties. First of all, the external conditions are variable. Anyone who has recorded the intensity of sunlight in measurements separated by several minutes is aware of this. The same is true for temperature and many other factors. Furthermore, these circumstances are *heterogeneous*, meaning that they differ in space. Places separated from each other by only a few decimeters can have entirely different climates. This phenomenon is known as "microclimate." For example, humus-rich soil is often acidic, yet fragments of shells, etc., can make the soil locally alkaline, such that the acidity level differs from centimeter to centimeter. Such measurements, when conducted in the field, show us the hopelessness of reaching a binding analysis, but can nonetheless be useful in certain cases when they delimit the boundaries of biological possibilities.

However, in all scientific observations it is important to be aware of variability (over time) and heterogeneity (in space).

While the field biologist speaks of "tamed creatures" in the laboratory and complains that laboratory methods are "unnatural," the experimenter has just as much right to reproach the field biologist for his apparent certainty gained by attempting to measure that which cannot be measured.

This contrast is not always as sharp as presented here, however, because whereas the immeasurability of various factors in certain environments (soil and atmosphere) is undeniable, this difficulty is not present – at least to the same extent – in other environments (particularly water).

An aqueous environment – be it bog, lake, or ocean – is certainly variable, but it is nonetheless much more homogeneous than other environments. Aquatic field biology is, perhaps for this reason, also much further developed than terrestrial field biology, and the biology of both fresh water (limnology) and salt water have long been sciences in which the question of "how" has often been answerable. Yet even here, the laboratory experiment must inspire.

The highest question a person can ask is "why." We ask this question in relation to the natural world around us in order to understand the appearance and behavior of organisms. This "why" is always causal and never goal-oriented.[3]

No matter how one analyzes vital functions in the laboratory, the organism is part of the Earth and its lot is interwoven with that of the Earth. Once again, in this context we must think of the enthusiast, he who opts for the out-of-doors. He is an "image seeker" and has perhaps been so since he was a boy. Later, in the laboratory, he becomes acquainted with experiments. Let him now return with confidence to the wilderness. Though aware of his limits and no longer so unbiased, he can test his knowledge on this natural environment. The Earth "as it is" remains the most important testing ground for our understanding of biology.

---

[3]Baas Becking seems to be warning of the difference between hypothesis testing and hypothesis proving.

This discourse is an attempt to describe the relationship between organisms and the Earth. The name "geobiology" simply expresses this relationship. This new word does not attempt to describe a new field. It rather tries to unite phenomena that have thus far been known to the different areas of biology as much as possible under *one* viewpoint.

I would like to thank the Board of the Diligentia Society, and particularly Dr. A. Schierbeek, for this opportunity they have offered me to organize my thoughts on this subject.

EDITOR'S NOTES

In this Introduction, Baas Becking highlights his view of the "geobiological" approach. He distinguishes between the field biologist (or naturalist) who is informed by observable biodiversity and patterns of species distribution, and the experimental biologist who puts the metabolic function of organisms into the context of laboratory controlled variations in environmental parameters. Finally, he argues that real insights ("how" and "why") come from combining these approaches so that the field biologist is informed by controlled and directed experiments on organismal metabolism and adaptation. This combined approach, which seeks to understand "the relationship between organisms and the Earth," is defined as "geobiology" by Baas Becking and bears much in common with the modern view. Baas Becking was modest in offering this definition and was quite specific that this definition "does not attempt to describe a new field." Little could he know that some 75 years later his "geobiology" is a thriving discipline of its own!

Baas Becking also highlights in this chapter the difficulty of placing an organism within an exact chemical and physical context in nature, particularly in terrestrial systems where chemical and physical gradients are large and "climates," as he calls them, are highly variable. Although our ability to determine small-scale variations in chemical and physical parameters (such as temperature, moisture, or oxygen) has advanced greatly since Baas Becking's time, understanding how organisms as individuals, or individual populations, interface with the chemical and physical environment remains a great challenge. For example, while we can measure in various ways the respiration rate of a terrestrial soil or a marine sediment, we still have a poor understanding of how individual members of the population contribute to this respiration. Part of the problem is that even now, we have difficulties in defining the true diversity of populations in nature, particularly microbial populations, and even for those members we can identify, we have difficulties in understanding their level of activity. This understanding, however, is beginning to expand with new approaches in molecular biology, including metagenomic sequencing for population diversity estimates as well as transcriptomic and proteomic approaches for elucidating the activity levels of individual populations in mixed microbial communities.

CHAPTER II

The Environment

Ein kleiner Ring
Begrenzt unser Leben.
Goethe

Imagine an ancient gate. Upon each column rests a sandstone ball, covered by a thin green layer of algae. In a hollow at the top lies a small pool of water, teeming with infusoria.[1] Now imagine a ball enlarged several million times: an enormous stone ball, the Earth, likewise with a thin green layer and a shallow pool, teeming with life. The Viennese geologist Suess named this layer the *biosphere*.[2] It is here, where atmosphere and lithosphere meet, that the highest known organizational form of matter has developed, closely related to – and in a certain sense the counterpart of – the Earth. It is the Earth itself, in its highest expression. This discourse is about this life, of and by the Earth.[3]

One cannot predict *a priori* the properties of molecules based on the properties of their atoms. This higher level of complexity brings with it new properties; the coordinated units together form a higher unit which has new properties. Nor can one predict the properties of the living state of matter based on its molecular configuration; life is a new property. But we can, *a posteriori*, test the properties of living matter against the conditions of its counterpart, inanimate nature: the counterpart, as the nest is to a bird, or as one shell to the other. Life without "counterlife" is unimaginable. This counterlife is the total

---

[1]This refers to all manner of small aquatic creatures, including ciliates, algae, bacteria, etc.
[2]Eduard Suess stated: "life is limited to a determined zone at the surface of the lithosphere. The plant, whose deep roots plunge into the soil to feed, and which at the same time rises into the air to breathe, is a good illustration of organic life in the region of interaction between the upper sphere and the lithosphere, and on the surface of continents it is possible to single out an independent biosphere." Suess, E. (1885–1908) *Das Antlitz der Erde*, F.Tempsky, Vienna.
[3]This expresses the concept of life as a geological force, as also eloquently described by Vladimir Vernadsky some eight years earlier in *Biosfera* [*The Biosphere*] (Vernadsky, V., 1926, Nauka, Leningrad). Baas Becking was apparently unaware of this, although he cited Vernadsky's *La Géochimie* from 1924 (120).

Baas Becking's: Geobiology, Or Introduction to Environmental Science, First Edition. Edited by Don E. Canfield.
© 2016 John Wiley & Sons, Ltd. Published 2016 by John Wiley & Sons, Ltd.

sum of external factors, in a given place and time, in which this life makes itself known. The word "environment" is used to describe this integration of factors (Claude Bernard, 21), as are "external environment" and *milieu cosmique*."[4] Geoffroy St. Hilaire calls it the *"monde ambiant."*[5] The geological concept of "facies" partially corresponds to this; the term refers to the sum of the primary properties of a given type of stone (52). In itself, the external environment is thus a geochemical concept, but this concept is empty without the life that is needed to fill it.

One could call the combination of these two concepts "geobiology," i.e., the biology that describes the relationship between living beings and the Earth. We therefore wish to consider life from the perspective of the Earth. Dr. M. Van Herwerden, who published a short piece entitled *The Developing Organism* earlier in this series of publications,[6] developed an entirely different – in many respects diametrically opposed – line of thinking. In her work, the internal environment is the dominant aspect, and while the motto above Chapter X of her work reads "The life of an organism is derived from its interior, not its exterior, and that inner world is framed by the external world,"[7] the motto of the present chapter should be "The development of an organism is derived from its interior, not its exterior. But its manifestation is guided by its exterior, not its interior, and the inner world is framed by the external world." The inner realization of an organism is not possible without an external environment that is conducive to that realization.

This means that any given form of life is only able to exist in certain specific environments. The spread of higher organisms over the Earth has been studied, and the boundaries (the so-called ranges) of the species have been established.

These boundaries enclose an area that can be thought of as the "natural environment".[8] For higher organisms, this environment apparently is not yet fully developed, but is constrained by, among other things, the limited possibilities for the organism to spread to new areas. In cases where humans have intervened in this process, it is possible to easily study the expansion of the natural environment. For example, the *Opuntia* (paddle cactus) and the rabbit were given perfect opportunities for expansion in Australia, as were the muskrat in Europe and the wild oat in California.

One could call the natural environment in its maximal extension the "global environment," and the difference between the two is determined by the possibility for expansion of the range (the natural area) of the species. In contrast with the other natural sciences, biology has until now been a purely

---

[4]"cosmic environment."
[5]"ambient world."
[6]Original Dutch title: *Het organisme in wording*. See Appendix for a list of the other titles in this series of scientific publications.
[7]Original German: "Das Leben der Organismen wird von Innen, nicht von Aussen hergeleitet und die Innenwelt ist in der Aussenwelt eingefasst."
[8]The "natural environment" refers to what we might now call "environmental range."

planetary, purely geocentric, discipline. Yet we can also easily apply Copernicus' example to biology and detach ourselves for the moment from the concept of "natural conditions." The Earth, in its landscapes, offers us only a limited number of possibilities, although there are scientists who, with some hesitation, accept the "artificial possibilities"[9] that the laboratory offers.

V. Vernadsky (120) writes: "An animal or plant created by a biologist is not a naturally existing body."[10] There even are biologists who see particular merit in "natural light," "natural soil," etc. The laboratory can now produce artificial possibilities that do not exist in nature. It is therefore possible to determine the conditions necessary for an organism's existence by developing cultures in the laboratory and comparing this area – the *potential environment* – with the natural environment. In doing so, one finds that the latter is never more expansive than the former. In most cases, the natural environment [of a given microorganism][11] makes up merely a small part of [its] potential environment,[12] in the same way that the natural environment of higher organisms is often just a small part of the global environment. The boundaries of the potential environment are not rigid, but can shift due to adaptation (7). It should also be noted that the influence of various environmental factors is not necessarily the same throughout all developmental stages of a given organism, adding a new complication. We will thus use three concepts in this discourse: the global, the natural, and the potential environments.

This discussion will be restricted to the so-called lower organisms, and we will only now and then mention a higher plant or higher animal. The literature – and the nomenclature in particular – about higher organisms is so extensive in the fields of geography, ecology, and so-called phytogeography,[13] that the behavior of higher organisms in relation to their environment can only be dealt with here in a cursory manner. Not only the vastness, but also the nature of the field, forces us to limit ourselves. For higher organisms, the potential environment is unknown in most cases, and so only the relation between the global and the natural environments is analyzed. In addition, one is confronted with complex concepts such as instincts and urges, as well as with an historic element that includes endemic species and relicts. These issues, too, must be left out of this discussion. Remarkably, by limiting the discourse to the lower organisms, their great and eternal relevance comes to light, as do some very general rules regarding their environment. In the future, we will need to make use of these rules. They have been consciously applied by our compatriot

---

[9]In the context of the text that follows, this seems to apply to the enrichment of populations from nature, which are thus taken from their natural environment. However, plants and animals altered by selective breeding might also be part of what Baas Becking is referring to.

[10]Original French: "*Un animal ou une plante de biologiste n'ést pas le corps naturel existant.*"

[11]Square brackets denote text that has been added by the translator or editor.

[12]This is an extremely astute observation on the metabolic flexibility of microbes in nature.

[13]Dutch: "*geobotanie.*"

M. W. Beyerinck[14] to his "enrichment method," and we could properly refer to them as Beyerinck's rules.[15] In 1913, Beyerinck wrote a treatise for the *Jaarboek van de Koninklijke Akademie van Wetenschappen* [Yearbook of the Royal Academy of Sciences] entitled *"De Infusies en de ontdekking der Bakteriën"* ["Infusions and the Discovery of Bacteria"]. In this treatise he discussed a new branch of science, micro-ecology, which developed precisely from the study of enrichments. What are the core concepts of this new science that describes the environment, the "house," the οικος of microbes?

The first rule can be formulated quite briefly, namely as "everything is everywhere." Microbial life is, in latent form, omnipresent. It is therefore cosmopolitan, universally distributed, merely waiting for an opportunity to manifest itself. A few examples will illustrate this.

In 1842, Goodsir found in the human stomach some peculiar "packet bacteria,"[16] which he named *Sarcina ventriculi*. W. F. R. Suringar described this organism in great detail in 1865. In 1905, Beyerinck found in garden soil a Sarcine which completely corresponded both in dimension and in shape with the stomach Sarcine. In 1911 Beyerinck succeeded in obtaining a Sarcine from the stomach of a patient in the Academic Hospital of Leiden which, as a culture, behaved completely the same and had the same dimensions as the form that he had isolated from garden soil. By cultivating it at 40 °C, in the absence of oxygen (anaerobic), with a high acidity level and with high concentrations of organic material (malt extract), he had thus created an environment that was specific enough to demonstrate the omnipresence of *Sarcina*.

As students we used to make excursions to find the curious purple bacteria. The work of C. B. van Niel, who described the environment of these organisms so precisely (91), has made it possible for one to now isolate these forms from almost any natural inoculum.

Further illustrations are not necessary. The confidence with which enrichment cultures are deployed, time and again, proves that this law is tacitly recognized. Beyerinck states in his above-mentioned article: "Experience has taught us that many germs[17] are so generally present in our environment that … often a precise answer can be given regarding the nature of the organisms which can or cannot be expected under certain living conditions …"

---

[14]Martinus Willem Beyerinck (or Beijerinck) (1851–1931) was the founder of the famous Delft school of microbiology. He was also one of the founders of virology as a field as well as the discoverer of nitrogen fixation and sulfate reduction as microbial processes. In the present context, he had an early understanding of the use of targeted enrichment cultures to understand the diversity and metabolic potential of microbes in nature.
[15]S. Winogradsky had already applied this methodology before Beyerinck, but our compatriot formulated very astutely the general principle that follows from that methodology. [Original footnote from Baas Becking.]
[16]Quotes are the editor's, as "packet bacteria" has no modern meaning.
[17]Throughout this chapter, the term "germs" (Dutch: *kiemen*) refers both to microbes that can reproduce under the right conditions and to cereal germs which, again, under the right conditions, can sprout to form new plants.

What expectation, then, is raised by the confidence that germs are omnipresent? Of course, in the first place, that we will be able to awaken them from their latent lives by creating a suitable environment, and if we wish to have a precise answer, in the sense of Beyerinck, this environment needs to be defined precisely. It must become a selective environment. And so we can add to the first rule stated here that *everything is everywhere*: but *the environment selects*.[18]

This is true both in nature and in the laboratory. The way in which the environment selects shall be the subject of the rest of this discourse. When nature or the researcher creates very definite conditions, the chance of a specific answer is also greater. A warm spring with blue-green algae, bacteria, and amoebae; a salt lake with its remarkable flagellates and crustaceans; as well as solutions containing no organic material that still always swarm with the same, specific life forms: these are illustrations of the second rule of Beyerinck! But field ecology can give us nowhere near a complete description of the environment; this privilege belongs to the laboratory.

In 1905, Söhngen (113) isolated bacteria that can live without organic material as long as they have methane, which they oxidize with the help of oxygen present in the air, using the energy released through this process to assimilate carbon dioxide, just as green plants do. In the year 1926 the author witnessed a massive growth of such bacteria in a water reservoir in southern California, where this water contained high concentrations of methane but no organic material, as often occurs in petroleum-rich areas. Nevertheless, in the pre-reservoir, where the sand settled out of the water being spouted up, there was a floating brown mass of bacteria which had rotted in the holding reservoir. These bacteria were methane bacteria [methanotrophs] that lived on this natural gas. Up to that time, methane bacteria had been known only as a laboratory product, but the natural environment offered here was specific enough to offer a specific answer!

The question thus arises as to whether there is relative certainty about whether by fabricating a specific environment one can necessarily capture its "counterpart."[19] Beyerinck himself writes about this in his above-mentioned discussion. Is it known that one can obtain an accumulation of lactic acid culture by adding a few drops of buttermilk to regular milk and then heating the mixture to 20–25 °C. "If one does the same experiment at a higher temperature, for example at 40–45 °C, then one can, for that matter in the same way, obtain other forms of lactic acid culture, such as koumiss,[20] kefir, yoghurt, leben,[21] *Matzoen*,[22] or how these drinks are further called in various countries.

---

[18]This is one of the most famous quotes in all of microbial ecology: "*Alles is overal*: maar *het milieu selecteert*."

[19]Presumably "counterpart" (Dutch: *tegenbeeld*) here means the organisms that would be expected to be of importance in the environments that the enrichment conditions were meant to reflect.

[20]Koumiss is a beverage of western and central Asia which is made from the fermented milk of a mare or camel.

[21]Kosher fermented milk.

[22]Could not find a translation (ed.)

These cultures nevertheless occur more rarely in our part of the world, such that in order to find them one must begin with larger quantities of them than above-mentioned ..."
There are rare and more common microbes. Perhaps extremely rare microbes exist: microbes whose distribution possibility is limited for one reason or another. Must we now assume, together with Ehrenberg, that the nature of the inhabitants of an enrichment are determined completely by chance? Although the salt factory in Boekelo has been in operation for more than twelve years now, in the summer of 1932 the author was unable to find any organisms specific to saline conditions there. Does this not lead us to the conclusion that microbes, and smaller organisms in general, also have *their* specific ranges?

No, because the environment selects. When studying an organism that rarely occurs in nature, we cannot assume, simply because it does not appear in an enrichment culture, that it therefore must have a specific range. This merely tells us that our description of the organism's environment is not yet sufficiently developed, not yet specific enough.[23]

The enrichments are certainly not determined completely by chance, although chance must play some role in it. Selectivity remains the principal driver. Beyerinck writes the following about this: "... Accurate observations teach that the nature of the inhabitants of enrichments remains different, depending on the living conditions realized in the enrichments."

If the rule "everything is everywhere" is really true, there must be a means through which germs are spread evenly over the Earth, giving rise to the old question of how germs are transported. We will further consider the most important means of transport to be transport through the atmosphere. Even before Pasteur had made his discoveries, organized particles, particularly grains of starch, had been found in the atmosphere. Ehrenberg also reports the presence of protozoa in dust. However, the honor of providing the inferential proof of the existence of airborne germs was Pasteur's, in his *Mémoire sur les corpuscules organisés qui existent dans l'atmosphère* ["Memoire on Organized Corpuscles which Exist in the Atmosphere"] (1861). In a very modern way, Pasteur showed that microbes of all sorts exist in the atmosphere, that their numbers are dependent on elevation, on topography and on the proximity of human activity (Figure II.1). The statement most important for us can be found at the start of his sixth chapter: "I believe I have rigorously established in previous chapters that all organized productions of infusoria[24] have no other origin than the solid particles which are carried by the air and constantly deposited on all objects."[25]

---

[23]This excellent point, raised by Baas Becking here, still preoccupies microbial ecologists.
[24]Meaning something like inocula in modern usage.
[25]Original French: "Je crois avoir établi rigoureusement dans les chapitres precedents que toutes les productions organisées des infusoires, n'ont d'autre origine que les particules solides que l'air charrie toujours et qu'il laisse constamment déposer sur tous les objets."

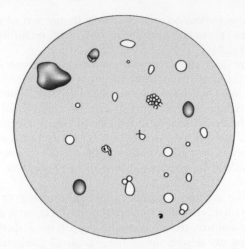

Figure II.1 Starch grains, yeasts, and spores, isolated from the atmosphere by Pasteur (from Pasteur).

We know that through these experiments Pasteur had already dealt the final blow to the experimental doctrine of *generatio spontanea*,[26] but it is more interesting here to note that Pasteur verified that a rain of germs falls from the atmosphere. The Viennese botanist H. Molisch has called these germs "aero-plankton," and in a discourse given in Vienna in 1916 (87), he gave a fine account of these airborne germs, in which one can also find the older literature about this subject. It appears that similar airborne plankton (which one can capture on glass plates spread with glycerine) includes: pollens, starch grains, mineral fragments, soot, spores, wood fibers, leaf fragments, linen, cotton, and wool. The cells that are able to germinate appear to belong to yeasts, molds, and bacteria (Figure II.2).

The cosmopolitan distribution possibility of small germs has already been explained. Keeping ourselves far removed from the otherwise stimulating speculations of Svante Arrhenius (4), we would like to take a closer look at a well-studied phenomenon: the spread of volcanic material.

On August 17, 1883,[27] the volcano Krakatoa erupted. On December 12, 1884, after a storm, Beyerinck (24) found crystals of kitchen salt as well as of an andesite-like mineral on his window pane (Figure II.3). This volcanic material, having been thrown far into the stratosphere, had needed a long time to return to the troposphere. A weak, light brown corona could be observed

---

[26]This refers to "spontaneous generation." Before Pasteur there was the common belief that microbes spontaneously generated in all of the strange places where they suddenly appeared.
[27]The major eruption of Krakatoa actually occurred on August 27, 1883.

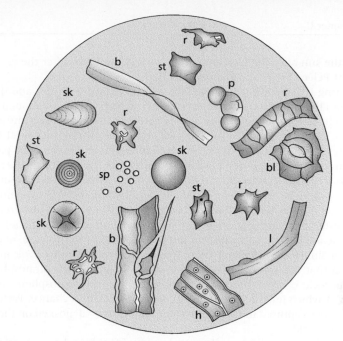

Figure II.2 "Airborne plankton." Starch grains, wool, linen, cotton, pollen, soot, and spores (from Molisch).

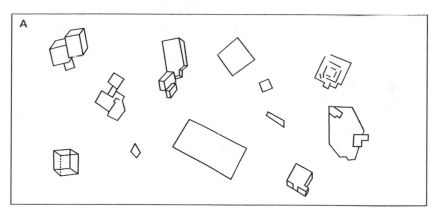

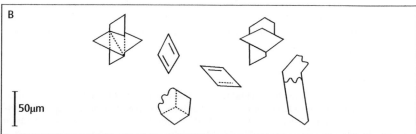

50μm

Figure II.3 From Beyerinck.

around the sun after the Krakatoa eruption and, similarly, after the eruptions of Mount Pelée (1902) and Katmai (1912).

This ring is known as the *Bishop's ring*. It has a width of 10°, and the distance from the sun to its outer edge is 23–23°.[28] With some limitation, one can calculate from this phenomenon how large the average diameter must be of the particles that cause it. This turned out to be approximately 1.85 μm. Humphreys (67) calculated the time it took after the eruption for the particles to return to the troposphere (11 km), assuming that the particles were spherical, with a specific gravity of 2.3. From this he concluded that it would take 1–3 years after the eruption before the material would be able to again reach the troposphere. When considering a dormant cell (cyst) of a protozoan, we must assume a specific gravity of 1.5 and a diameter of approximately 5 μm. When one applies the comparison used by Humphreys to this case, one finds speeds for similar particles that are 6–7 times as great as those of volcanic ash, assuming that the cysts are catapulted 40 km high into the stratosphere. A similar particle would remain two to six months in the stratosphere, and the location at which it once again reaches the Earth is left to chance. Particles of this size, once captured in the stratosphere, can thus be deposited on the other side of the world!

Also in the troposphere itself, as we know, material can be transported several hundreds of kilometers. Pine pollen, the famous sulfur rain, is a striking example of this, and the spread of microbes gives indirect evidence of this. The German plant geographer Walter (123) considers organisms such as bacteria, molds, and planktonic algae also to be cosmopolites, and states the most important source of their distribution to be the wind.

As long ago as 1884, Géza Entz Sr. (50) had explained the presence of the unusual, partially "marine" protozoan population of the salt lakes in Transylvania by presuming that they were carried by the sea wind. "This hypothesis is the only means by which I can explain the interesting fact of the presence of exquisite marine (littoral and brackish water) infusoria in continental saline lakes and ponds so far from the sea, as especially found in Transylvania."[29] Mrazek made a precise formulation of the same in 1902 (89), prompted by his study on the fauna of greenhouses; the animals truly exist there, where their requirements for life are met. The "dissemination of germs" is so massive that they will be present everywhere, but selection takes place and only the forms adapted to the given circumstances will appear. Mrazek blames (probably correctly) the insufficient knowledge about the spread of the lower animals more on insufficient interest than on the lack of these organisms in given places. Springtails, *Acaridae*, many arachnids, nematodes, crustaceans, protozoans, *Tardigrada*, and rotifers fall into this group, as do most

---

[28]This is apparently a typo in the original text. Perhaps he meant 23–28 degrees?

[29]*"Nur durch diese Annahme kann ich mir die interessante Tatsache erklären, das exquisit marine (litorale und Brackwasser) Infusoriën in weit vom Meere entfernten kontinentalen Kochsalzteichen und Tümpeln, wie namentlich in Siebenbürgen, vorkommen."*

freshwater algae, which continually tend to produce the same familiar forms when mass-produced. One can demonstrate with very simple experiments that it is not infectious material – be it ditch water, garden soil, tannin, or sphagnum – but rather the applied culture liquid that determines in the first case which organisms will appear. Gistl (55), who recently made a study of soil algae, also comes to this conclusion and says: "The sum of the chemical and physical conditions prevailing in a given soil are decisive for the composition of the algal flora."[30]

Experiments carried out by Sinia in the Botanical Laboratory of Leiden show that the soil algae in Meyendel are, in latent form, practically homogeneously distributed, and that one can continue to isolate the same forms by means of selective culture mediums. For example, green algae [of the genus] *Hormidium* are present in every substrate, whereas blue-green algae [of the genus] *Fischerella* occur only in soils containing no calcium or nitrogen, because blue-green algae develop in nutrient solutions with no fixed nitrogen. By varying the nutrient solutions, the presence of a large number of forms previously unknown as soil algae was able to be demonstrated.

Many biologists would disagree with the purport of this argument (Donat, 45; Florentin, 54). There are many who establish the ranges of microscopic organisms, such as freshwater algae, and consider the forms found in our peat bogs to be relicts from the ice age. Two factors can perhaps explain this opinion. In the first place, the insufficient knowledge we have regarding the spread of the organisms, and second, our insufficient knowledge about the potential environment of these organisms. There are indeed organisms for which the environment is so specific that they can survive in only a few places on Earth. These locations can be separated from each other by great distances, suggesting a geographic correlation that in actuality does not exist. It is, however, remarkable that Charles Darwin, in his *Origin of Species*, recognizes the great importance of cosmopolitan distribution and of the selective environment, and mentions only in the second place the spread of organisms not based on these factors (and which yet may be considered one of the foundations of his theory). He then says in Chapter XII: "In considering the distribution of organic beings over the face of the globe, the first great fact which strikes us is, that neither the similarity nor the dissimilarity of the inhabitants of various regions can be w h o l l y[31] accounted for by climate and other physical conditions."

With even more confidence – and this is actually the matter that interests us most here – we can say that the organisms that fulfill an important, specific function in the material cycle, in the geochemical context, are omnipresent in latent form, and that their presence in this vegetative form is caused by the nature of the external circumstances. Because of this, there sometimes exist, to

---

[30]"*Die Summe der chemisch-physikalischen Bedingungen die im gegebenen Boden herrschen, sind Ausschlaggebend für die Zusammensetzungder Algenflora.*"
[31]Emphasis is placed on "wholly" through the use of spacing by Baas Becking.

use the terminology of Beyerinck, "naturally pure cultures" (the literature written on this subject up to 1907 was summarized by Ferd. Stockhausen, 116). Instinctively, and often with great keenness, the field biologist recognizes a specific environment: the identified constant of seawater, the warm freshwater pool, the acidic peat bog, or the alkaline salt lake. But the environment itself is again an integration of factors, and the analysis of this complexity places us anew in the midst of the "*Chaos Infusorium*," as Linnaeus used to call the still incipient microbiology!

However, order can be brought to the chaos. In this regard we instinctively first notice the colored organisms, those which therefore absorb and often also utilize visible light. Some organisms thrive in the exclusion of oxygen, while others require it. Some need organic material, others can build their bodies out of inorganic material. The preparation of the "laboratory environment," which must capture a specific form from out of a large number of organisms, is predicated on these and similar characteristics. Very briefly, we may state some of the principles of this method. Green plants, in addition to several groups of colored bacteria (green and purple bacteria) are capable of converting the energy contained in sunlight into chemical energy, with which they then convert the carbon dioxide found in water or air into complex substances (sugars, for example). One calls these organisms *photosynthetic*; they need no organic material, as they make this themselves.

Winogradsky (127) has now discovered the existence of colorless organisms which can do without organic substances for their nutrition because they derive the energy needed for their metabolic processes from chemical reactions between inorganic materials in the environment. One calls this way of life *chemosynthetic*. Such bacteria utilize the oxidation of ammonium ions into nitrite ions. One categorizes photosynthetic and chemosynthetic organisms as *autotrophs*, meaning that these beings feed themselves and are not dependent on others either directly or indirectly. All other organisms (including ourselves) are known as *heterotrophs*.

In the table below, several factors concerning the specific environment have been summarized. The usefulness of the table can be illustrated as follows. Suppose that one wants to culture green bacteria. We find these organisms in the table and quickly find the combination of factors required in their environment. This is: an environment rich in H_2S, a weakly acidic solution, light, and anaerobiosis. The other way round, one can ask what would happen if one were to make a nutrient solution under aerobic conditions (i.e., allowing the presence of free oxygen), under autotrophic conditions (i.e., without the presence of organic material) and to which ferrous iron is added. According to Winogradsky, one could then expect iron bacteria, since the oxidation from ferrous to ferric iron produces enough energy to allow these organisms to assimilate carbon dioxide.

The table will be examined more extensively in the chapter which covers the material cycle. Various environmental factors mentioned above, such as solar radiation, temperature, and chemical influences, will be discussed in the following chapters.

Schematic overview of some environmental factors of microorganisms					
Aerobic	Heterotrophic		Basic		Many bacteria
			Acidic		Many fungi
			N nonexistent		*Azotobacter*
	Autotrophic		Fe^{2+}		Ferrobacteria
			NO_2		*Nitrobacter*
			NH_4^+		*Nitrosomonas*
			CH_4		Methane bacteria
			CO		*B. oligocarbophilus*
			H_2		Hydrogen bacteria
			S		*Thiobacteria* (Waksman)
			H_2S – T high		*Thiobacteria* (Miyoshi)
			H_2S – T low		*Thiobacteria* (Winogradsky)
Anaerobic	Dark	Heterotrophic	N nutrient base	T high	*Sarcina ventriculi*
				T high or low	*Microspira desulfuricans*
				T low	Beer yeast, *Clostridium butyricum*
			High nitrate level		*Micrococcus denitrificans*
			N nonexistent		*Clostridium pasteuranium*
			H_2X		purple bacteria (Muller)
	Light	Autotrophic	H_2S	basic	little H_2S	purple bacteria (van Niel)
				acidic	much H_2S	purple bacteria
						green bacteria
			H_2O	N nonexistent		blue-green algae
				NH_4^+	basic	Diatoms (partially)
					acidic	Desmidiaceae
				NO_3^-	saline	Dunaliella
					basic	Cladophora
					acidic	Protococcales
Cosmopolites with a very broad range					Sulfate reducers, blue-green algae, amoebae

EDITOR'S NOTES

Baas Becking was a product of the famous Delft school of microbiology. He studied under the guidance of M. V. Beyerinck, who, among many pivotal contributions, systematized the approach of establishing enrichment cultures as a way to isolate members of a natural population with specific metabolic properties. This method, still a cornerstone of microbial ecology today, involves using media of specific compositions, and under specific conditions of light, temperature, and pressure, to isolate members of a microbial community adapted to these conditions. This approach led to the discovery of countless microbial populations, many of which were otherwise difficult to detect or were undetectable in the natural environment where they lived. This approach also led to several key observations. One is that the diversity of microbial populations in a natural environment is large. Another is that microbes living in an environment tend to be well adapted metabolically to the geochemical conditions offered by the environment. A third is that after careful study of the isolated microbes, their metabolic potential typically exceeds what is necessary to live in the environment from which they were isolated.

These observations, especially the way that microbes adapt to the environment where they live, combined with the widespread global distribution of similar microbial species, led Baas Becking to highlight what he called Beyerinck's rules. He summarized these rules into what stands as one of the most recognizable expressions in microbial ecology today: that is, *"everything is everywhere*: but *the environment selects."* Decades of subsequent study have only reinforced this notion.

The focus on the relationship between microbe and environment, as revealed through enrichment culture studies, is also apparently what inspired Baas Becking to suggest geobiology as a new field of endeavor. In this chapter he defines geobiology as "the biology that describes the relationship between living beings and the Earth." This remains a valid definition today. The same focus, on the relationship between between microbe and environment, reveals the remarkable range of microbial metabolisms in nature, and the importance of microbes in the chemical transformations of matter. This led Baas Becking to offer the remarkable statement: "by limiting the discourse to the lower organisms, their great and eternal relevance comes to light." Most modern microbial ecologists would fully embrace this notion.

Overall, this chapter is very modern, although written more than 80 years ago. Most of what Baas Becking discusses remains not only true, but central to microbial ecology and geobiology as fields. We have, however, learned many things in the subsequent decades. The extraction and analysis of DNA and RNA from the environment has revealed a far greater microbial diversity than Baas Becking knew, or perhaps even could have guessed. We have learned that the enrichment technique often yields the fast-growing "weeds" in the system, but not usually the dominant species within an environment. However, careful consideration of media composition and culture conditions is beginning to yield many of the major microbial species. We have also learned that in many ecosystems the microbial populations are narrowly adapted to the environment where they live. An example is the distribution of the cyanobacteria *Prochlorococcus* sp. in the oceans, whose individual "ecotypes" are restricted to a narrow environmental range of light and nutrient availability. I'm quite certain that Baas Becking would have been delighted to see these developments in geobiology and microbial ecology, and happy to see that much of what he wrote in the 1930s remains valid today.

CHAPTER III

Environmental Factors: Solar Radiation

Almost all of the radiation that affects the physiology of life on Earth originates from the Sun. In addition to the indirect geochemical or geophysical effect this radiation has on organisms, we also know of a number of processes which are caused directly by sunlight. One such example is *sight*, the activation of various substances by light (e.g., ergosterol by ultraviolet light), but especially the absorption of light by colored plant cells, where the absorbed light – transformed into chemical energy – can be of benefit to other, colorless cells. As this process of photosynthesis is crucial to the maintenance of life on Earth, it is therefore also of utmost importance that the quantity and quality of this light is researched in a natural environment.

Amount of radiation

Through meticulous measurements, Abbot has been able to ascertain that the intensity of sunlight beyond Earth's atmosphere amounts to approximately 1.94 calories per cm^2 per minute (solar constant). At our latitude,[1] at most around 70% of this reaches the Earth's surface; the rest is absorbed by the atmosphere, reflected, or scattered. The atmosphere mainly scatters short-wavelength light; the degree of scattering is inversely proportional to the fourth power of the wavelength. The scattered short-wave light is what causes the sky to appear blue.

The amount of solar radiation that reaches the Earth depends on, among other things, the distance to the Sun and the angle at which sunlight hits the Earth. If the total daily solar radiation at the equator on the date of March 20, with clear skies, is equal to 1.000, then on the following dates, at a latitude of 50° this will be:

March 20   = 0.633
June 21   = 1.154
September 22 = 0.625
December 21   = 0.206

---

[1]Between 51° N and 54° N for the Netherlands.

Baas Becking's: Geobiology, Or Introduction to Environmental Science, First Edition. Edited by Don E. Canfield.
© 2016 John Wiley & Sons, Ltd. Published 2016 by John Wiley & Sons, Ltd.

The vertical intensity of the direct solar radiation $(I) = I \times \cos d$, in which d is the zenith distance in degrees.

Aldrich determined that with clear skies, 27% of direct solar radiation is indirect solar radiation (1). This percentage increases with cloudy skies and when indoors.

Nature of solar radiation

As a glowing body with a temperature of approximately 6500 °C, the Sun emits radiation in a fairly continuous spectrum. However, in the atmosphere, oxygen and ozone absorb waves with a length of less than 300 nm (1 nm = 10^{-9} m), while water vapor and carbon dioxide mainly weaken the rays of light that have a longer wavelength (Figure III.1).

The percentage of sunlight that reaches the Earth's surface at our latitude has been calculated by Abbot and Fowle (2).

Wavelength in nm	Beyond the atmosphere (random units)	% penetrating
390	4614	44.5
430	5321	60.0
450	6027	64.0
470	6240	67.1
500	6062	70.5
550	5623	73.9
600	5042	76.0
700	3644	83.9
800	2665	86.5
1000	1657	90.1
1600	532	93.0
2500	43	87.0

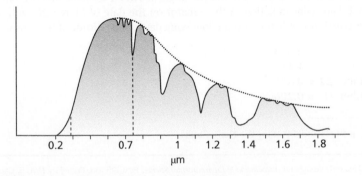

Figure III.1 Intensity of the solar spectrum (from Langley).

Beyond the atmosphere there is a maximum in the blue-green spectrum and on Earth a maximum ranging from yellow-green to red, depending on the thickness and nature of the atmosphere (e.g., evening sun). The distribution of light in a certain landscape is very complicated. In addition to the above-mentioned factors, one must also take into account the optical traits of leaves: their reflecting, absorbing, and transmitting abilities, as well as the placement of the leaves (leaf mosaic), which determines the so-called "environmental light" as in a forest (Seybold, Walter). In the short scope of this book we will limit ourselves to a more homogeneous environment, namely water. The distribution of light in water has been studied repeatedly and with the use of various methods (Shelford and Gail, 111; Birge and Juday, 27; Atkins and Poole, 6). Many biologically interesting facts have been discovered in these and other studies; a few of these facts are mentioned below.

a *Reflection, refraction, and absorption*
At the interface of air and water the amount of light reflected differs depending on the zenith distance of the sun, the cloud coverage, the nature of the water surface and the color of the water. Only 20–80% of the light is let through and refracted (the refraction index for fresh water at 15 °C = 1.334, for seawater at 15 °C = 1.338), as a result of which the rays descend at a steeper angle. Due to the refraction, even light that enters almost horizontally is able to enter the water. Non-homogeneity in temperature and salinity causes an even greater shift of the refraction index in water; the dispersion of light also plays a role in this.

The scattering of light, which takes place on Earth as well as in the atmosphere, is also extremely important. Because of this scattering it is possible still to perceive light in a horizontal direction underwater, as even light approaching horizontally would only make an angle of 48° with the vertical axis. In practice, however, it is mainly the suspended particles (air, plankton, mud) that influence the dispersion of light in water. Atkins and Poole (6) discovered that in clear water, 20% of the light present at the surface could still be found at a depth of 10 meters, while in muddy water only 0.03% was found.

The transmission of total solar radiation through the water in the harbor of Nieuwediep, not counting loss through reflection, is depicted in Figure III.2 (determined through the use of a blackened thermo-element, a so-called solarimeter). Figure III.3 provides several results as found by Atkins, obtained by using photoelectric cells. Comparing the two figures, it is clear that the water in Nieuwediep has extremely limited transmission capacity, which is caused by the large quantity of suspended mud.

b There is one more very important factor that can be added to the above-mentioned factors: *the specific light-absorbing capacity of water*. It is generally known that deep water has a blue color. This means that blue light passes through water quite easily. The adjacent ultraviolet also easily passes through fresh water, whereas seawater absorbs this light quite strongly. This strong absorptive quality most likely has to do with the high level of magnesium (Hulburt, 65). Water also strongly absorbs infrared light.

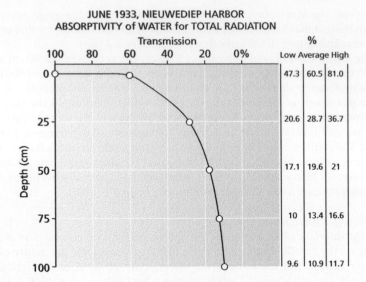

Figure III.2 June 1933, Nieuwediep Harbor. Transmission capacity of water for total radiation.

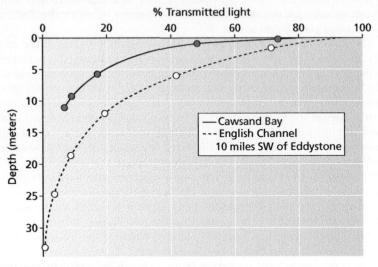

Figure III.3 From Poole and Atkins.

The specific absorptive qualities of water cause several curious phenomena. Seen from the surface, red objects located in deep water appear blue. When these objects are retrieved, they change color as they are brought to the surface. In June of 1930, 10 miles south of Bermuda, William Beebe was lowered into the sea in a metal ball, to a depth of 1400 feet. At 800 feet the color blue was almost monochromatic: a narrow zone around 500 nm. He remarked, on the light that reached him at 1400 feet: "At 1400 feet the outside world seemed born of a single vibration: blue, blue, forever and forever blue." Beebe was only able to observe in a horizontal direction and was therefore only able to detect scattered light (Hulburt 66). Not only the selective absorption, but also the scattering of light contributes to the blue color of the oceans.

As is known from physics, monochromatic light is absorbed by homogeneous media in accordance to the Beer–Lambert law, with the following formula:

$$I = I_o e^{-kcd}$$

in which:
$I =$ intensity of the transmitted light
$I_o =$ intensity of the incident light
$e =$ base of the natural logarithm
$c =$ concentration (of substance in a solution)
$d =$ distance the light travels through the material (path length)
$k =$ constant

In seawater the concentration [of salts] is constant, but the incoming light is white; however, the rule applies only to monochromatic light; this means that different rules apply to daylight (Bouguer). Most of the infrared light is absorbed in the upper layers, while the blue is much less muted. Absorption of daylight by water must therefore follow a more complicated rule. Figure III.4 displays the transmission of various wavelengths in seawater, data which can also be deduced from the following table.

λ in nm	k (d in centimeters)	
400	9.32×10^{-4}	
420	4.14	
440	2.30	
460	1.66	
480	1.54	← maximum transmission
500	1.61	
520	1.95	
540	2.98	
560	4.14	
580	7.46	
600	16.5	

c *Green plants*

In the section above, several factors that can influence the nature and the intensity of light in water were very briefly discussed. The distribution of green plants under water depends on these factors. Generally, in even the clearest of waters, these plants cannot be found deeper than approximately

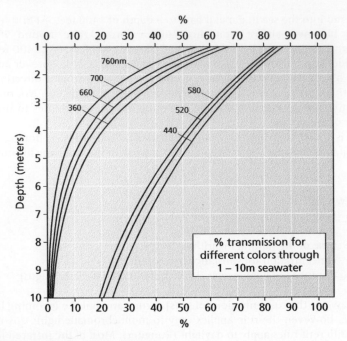

Figure III.4 From data by Atkins, 1932.

180 m. Although calcareous coralline algae (*Lithothamnion*) have been found at deeper levels, the assumption can be made that these organisms do not use photosynthesis but are saprophytes.[2] Comparing Figures III.2 and III.3 shows how variable the transmission of light within water can be, and a green plant which does not have any organic material available to it will not be able to exist below a certain level of light intensity, as at that level (and at a certain temperature) the processes of respiration, photosynthesis, catabolism, and anabolism balance each other out. This is called the "compensation point." For green algae, especially in the murky North Sea, this point lies just below the surface, as depicted in Figure III.5.

With light, the green alga *Ulva lactuca* L. is able to produce a large amount of oxygen in a short time. This can be assessed by using the Winkler test for dissolved oxygen. After adding an amount of iodide, a certain amount of iodine will be released; this can be titrated with a solution of sodium thiosulfate and provides a measure for the amount of oxygen present in the water. Figure III.5 shows the amount of oxygen (in 0.1 mL normal sodium thiosulfate solution per 100 mL water) that the seaweed *Ulva* (9 grams per liter) produced within two hours on a sunny day, at various depths in the Nieuwediep Harbor

---

[2]An antiquated term which can be taken to mean non-photosynthetic plants. We now know that these organisms are animals, and at these depths they lack any photosynthetic partners.

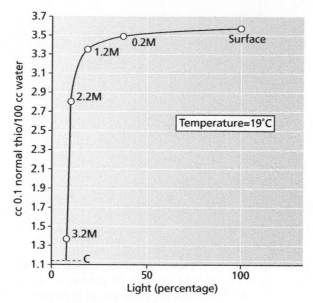

Figure III.5 Excess oxygen and amount of light (measurements by Ms. J. Ruinen and J. Zaneveld).

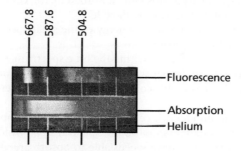

Fluorescence

Absorption

Helium

Figure III.6 Spectra of *Hormidium flaccidum*.

(August 1933). The point C on the figure is the amount of oxygen in seawater without seaweed; this point also represents the compensation point.

For *Ulva* the compensation point appears to be at a depth of approximately 3 meters. In clear water and in midsummer the compensation point can lie much deeper; depths of over 20 meters have been measured.

The light-absorbing capacity of green cells is mainly caused by chlorophyll, which contains magnesium, together with a hydrocarbon, carotene, and the carotenoid lutein. In Figure III.9, the transmittance of a living green cell (*Euglena*) is shown for the various spectral colors.

Figure III.6 is an image of the absorption spectrum of a living multicellular, non-branching filamentous charophyte green alga (*Hormidium flaccidum*). This

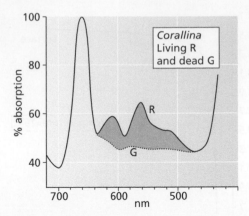

Figure III.7 Transmittance of *Corallina*.

figure also shows the fluorescent spectrum of the alga. If one exposes a green cell to light, it will emit a red light. As Molisch phrases it: the green cell is a "factory of red light." So if radiation is absorbed, at least part of this will be transformed into red light with a set limit of wavelengths.

As water is a specific absorber, the useful effect of photosynthesis will be influenced by this specificity, if this light, absorbed by the colored plant, is to be utilized for the assimilation of carbon dioxide.

As we saw earlier, the maximum energy of the solar spectrum usually lies within the green-yellow spectrum.

The transmittance of water lies within the blue spectrum, so blue and green are the dominant colors in deeper waters. The green light cannot be utilized by the green plant, as it "leaks" this color; the color green passes through practically unchanged.

At the end of the nineteenth century, W. Engelman proposed the theory that colored seaweeds had adapted to the spectral light which reaches them at a given depth, with only those weeds able to absorb green light appearing at greater depths (passive chromatic adaptation). Red algae and blue-green algae possess so-called "accessory pigments," extra colors in addition to the green (chlorophyll) and the orange (carotene and lutein) which occur in almost all green cells. Figure III.7 shows the absorption spectrum for the red coralline alga *Corallina*, where the upper line is obtained by determining the absorption of various colors by the living alga. The lower line shows the absorption of an alga killed with chloroform and then extracted by water.[3] While there are no noticeable differences between the living and the dead *Corallina* in their absorption of red and blue light, the dead weed's absorption of green light is far less than that of its living counterpart, and its aqueous

---

[3] Gas-liquid extraction.

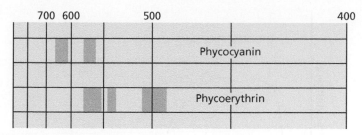

Figure III.8 From Kostychev, 76.

extract has a reddish-purple hue. This extract contains two protein-like materials, phycoerythrin and phycocyanin, of which the colored component is related to bile pigments (bilirubin). These pigments occur in both red algae and blue-green algae. Figure III.8 depicts the spectrum of an aqueous solution of both pigments.

This shows that red algae are indeed better able than the blue-green algae to utilize the light available at greater depths. However, in shallow waters one cannot find much evidence of the formation of light-based zones. Engelmann wanted to stretch his theory to include brown algae, which contain the accessory pigment fucoxanthin (which can be seen as an oxidation number of carotene). However, as matters here are not as clear as with red algae, we will satisfy ourselves with just this short annotation.

Buder (33) pointed out a different phenomenon which could possibly also be understood as passive complementary adaptation. Purple bacteria, which are also photosynthetic, are often found under a layer of green algae or duckweed. These organisms must therefore be able to utilize the light that is let through by the green cells above them. The absorption band widths of the chlorophyll and of the pigmentation of the purple bacteria are more or less complementary.

This becomes even more apparent when one looks at a graph depicting the absorption capacity for different colors by green as well as purple organisms (Figure III.9). The complementary character of the two absorption curves is very clear.

Examples of *active* chromatic adaptation are mainly observed with blue-green algae. Some forms are naturally able, in certain circumstances, to take on a color which is more or less complementary to that of the available light.

In the beginning of this chapter the focus was on the importance of water's absorption of the neighboring infrared spectrum. This type of solar radiation is already absorbed by shallow layers of water, whereas it is known that household salt allows this radiation to pass through easily. The distribution of purple bacteria (only in shallow waters, but under thick layers of salt, see Chapter X) and the important absorptive qualities they have in the adjacent infrared, as well as the fact that said organisms, when irradiated by a spectrum, accumulate within this infrared range, leads one to the presumption that purple bacteria are possibly able to utilize the infrared light for their carbon dioxide assimilation.

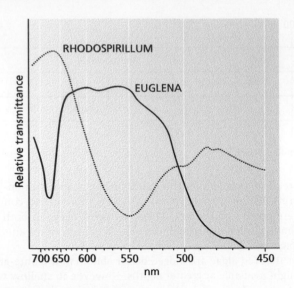

Figure III.9 Transmittance of green and purple cells.

In this chapter, light has only been discussed as a source of kinetic energy for the photosynthetic cell; yet the number of physiological processes influenced by light is of course much larger. In geochemical terms, however, the "capture" of energy remains the most important for our purposes, and other photochemical processes, however fascinating, will not be discussed within the scope of this book. The efficiency of carbon dioxide assimilation in various environments will be discussed in Chapters VII, VIII, and IX.

EDITOR'S NOTES

While Baas Becking was raised on the traditions of the Delft school of microbiology, he was trained as a botanist and was keenly interested in light and its availability to plants. Indeed, he also wrote a separate book on *Radiation and Vital Phenomena* (1921, Pranava Books, Delhi; a product of his PhD) in which he rigorously explored the energetics of light utilization by plants. Baas Becking's main concern in the present chapter is to discuss the availability of light to photosynthetic organisms in aquatic environments. This chapter includes a discussion of how both the intensity of light and its spectral properties are attenuated in different environments and how photosynthetic organisms adapt to the available light. In a particularly insightful analysis, Baas Becking relates how populations of purple bacteria living under populations of green algae and other plants must have absorption spectra complementary to (and not the same as) those for the algae and plants.

He continues with the observation that purple bacteria can live at reasonably great depth in encrusting salt, where the infrared (IR) spectrum of light can be transmitted. He speculates that the purple bacteria have pigments able to utilize IR light for photosynthesis, which turns out to be quite true.

Baas Becking wrote this chapter when the measurement of light and its spectral properties in aquatic environments was a new and emerging possibility. Since that time, there has been much focus on measuring the quality of light in aquatic environments, reinforcing much of what Baas Becking discusses here. Notable advancements, particularly as they relate to prokaryote photosynthesis, include the development of microsensors to measure light quality in highly stratified microbial communities with a depth resolution of tens of microns. These studies have reinforced what Baas Becking discussed regarding the complementary nature of the light-absorbing properties of stratified photosynthetic communities. Unforeseen by Baas Becking was the presence of exceptionally low-light-adapted sulfide-oxidizing phototrophs in stratified basins such as the Black Sea. These phototrophs can photosynthesize with as little as 100,000 times less light than ambient at the water surface. Also unknown to Baas Becking were the low-light-adapted oxygenic phototrophic communities in the deep ocean where light meets nutrients. These low-light-adapted populations are active photosynthesizers, often generating a deep Chl a maximum layer.

CHAPTER IV
Environmental Factors: Temperature

> ... organisms exist
> only within a few degrees of the long scale
> rangeing from measured zero to unimagin'd heat ...
> *Robert Bridges*

(1) Water

In organisms whose internal and external environments are both aqueous, a clear description of the influence of temperature upon these life forms is shown to be most closely related to the thermal properties of water. This is also extremely evident with "terrestrial organisms," and it would be relevant to take a closer look at a few properties of water that are influenced by temperature.

a *Solubility of gases*

The solubility of gases in water will be discussed in Chapter V. This solubility decreases when temperatures increase, which is why extremely low levels of oxygen are found in hot springs. The carbon dioxide – carbonic acid equilibrium also shifts, resulting in decreasing levels of available carbon dioxide and bicarbonate.

b *Viscosity*

The viscosity of water decreases fairly quickly with increasing temperature. Where the ability of organisms (plankton) to float in water is dependent upon the viscosity of the water as well as on the relation between the organism's surface area and its volume, one (Steuer 114) has often explained the large specific surface of organisms living in warm water, as opposed to those living in cold water, as an adaptation to the viscosity of the water.

Figure IV.1 depicts the influence of temperature on viscosity for different types of water, from which we can see that salinity is another factor that greatly determines this property.

Baas Becking's: Geobiology, Or Introduction to Environmental Science, First Edition. Edited by Don E. Canfield.
© 2016 John Wiley & Sons, Ltd. Published 2016 by John Wiley & Sons, Ltd.

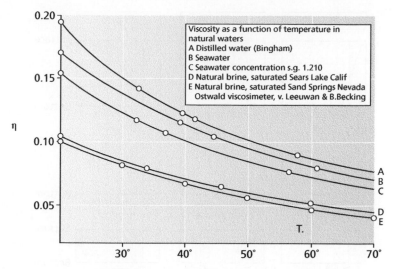

Figure IV.1 Viscosity of various types of water, shown as a function of temperature.

Viscosity comes into play with many other functions of organisms. Bacteria in highly saline waters often become active only at fairly high temperatures (45 °C), whereas analogous forms in fresh water are already active at 20 °C. It is premature to say that the bacteria present in salt lakes are thermophiles, as long as we do not know to what extent the viscosity of the external environment slows down their movement.

Viscosity is also very important for the internal environment. The innermost layer of protoplasm (the so-called endoplasm) can flow in many plant cells. The speed of this flow increases with higher temperature. According to Hille Ris Lambers (60), the speed of the flowing protoplasm is a function of the protoplasm's viscosity. However, more complicated laws (12) apply for the outer layer of the protoplasm.

c *Density*

Pure water has maximum density at 3.96 °C. This maximum, which is probably caused by the combined nature of the water molecules (see Chapter V), shifts to lower and lower temperatures when more and more salt is dissolved in the water, such that it [eventually] lies below the freezing point of seawater.

The following table is taken from Dittmar (42). The solution is compared with a standard liquid at 15 °C.

	0°	5°	10°	15°	20°	25°	30 °C
Concentration ↓	1.00077	1.00087	1.00060	1.00000	0.99911	0.99796	0.99659
	1.01130	1.01120	1.01075	1.01000	1.00898	1.00774	1.00630
	1.02182	1.02152	1.02090	1.02000	1.01886	1.01751	1.01600
	1.03228	1.03179	1.03102	1.03000	1.02876	1.03732	1.02572

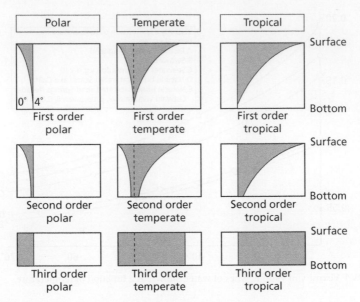

Figure IV.2 Classification of freshwater lakes (from Whipple).

One can clearly understand from this table that when a volume of fresh water cools from 15 °C, its density will first increase until it reaches approximately 4 °C. With a further drop in temperature, the density of the water once again decreases. The phenomenon of the so-called spring and autumn inversion in deep lakes rests on this property.

Just as weather conditions play out mainly in the troposphere, while the stratosphere lies serene and undisturbed above it, one finds that the upper layer [of a lake] is influenced by weather and wind (the epilimnion), while a mass of water that is hardly touched by atmospheric influences (the hypolimnion) lies beneath it.

The peculiarities of these layers will be further discussed in Chapter VIII; here I would just like to mention that in extreme cases, the hypolimnion can reach a temperature of 4 °C. Figure IV.2, taken from Whipple (126), shows the variations in temperature in the epilimnion and the hypolimnion in different types of lakes. The top set of diagrams shows the lakes "of the first order," by which Whipple means masses of water in which, because of the great depth, the hypolimnion is fully developed. The middle diagram of the top row depicts a deep lake in a temperate climate. The uppermost horizontal line shows the seasonal temperature fluctuation in the epilimnion – the upper water layer – while the bottom horizontal line shows the temperature of the hypolimnion (4 °C). Twice a year, the temperature of the epilimnion is the same as that of the hypolimnion. One of these times is in the spring, when the ice has melted and the temperature in the upper layers of the water slowly rises to 4 °C. When the temperature of the entire lake reaches

4 °C, the spring inversion takes place. Tiny differences in temperature then result in huge convection effects, and the entire lake is "stirred up" in a vertical direction (spring inversion). A similar process takes place in autumn. Needless to say, these inversions bring with them enormous biological consequences. The interpretation of the other figures is left to the reader. Lakes of the second order have an incompletely developed hypolimnion, while lakes of the third order show no stratification, due to their shallowness. The word "tropical" in the third column should not be taken too literally, seeing as how the majority of tropical lakes seldom if ever have a temperature of 4 °C in the epilimnion.

In the oceans, where, in addition to temperature, salinity level determines the specific gravity of the water, epi- and hypolimnions[1] can also appear under certain circumstances. Here, the phenomena are much more complicated than in fresh water. We therefore refer to specialist works (e.g., Harvey, 58) for this matter.

d *Other thermal properties of water*
Specific heat, latent heat of vaporization, heat of fusion, and thermal conductivity are all properties that exercise an influence upon the environment. Water's high specific heat and its high heat of vaporization are praised by, among others, Henderson (59), who describes an aqueous environment as one big thermostat, an apparatus of constant temperature. Much heat can be absorbed and given off without the temperature changing noticeably, whereas when the temperature rises, more water evaporates, and through this evaporation heat is absorbed, such that here again a "regulating" influence is noticeable. This is true not only for the external environment, but also for the organism itself, which consists largely of water.

> It is therefore incontestable that the unusually high specific heat of water tends automatically and in a most marked degree to regulate the temperature of the whole environment, of both air and water, land and sea, and that of the living organism itself (59, p. 91).

This is not the place to delve further into the properties of water. Solubility of gases, viscosity, density, and specific heat are mentioned here only to show that the conclusions, derived here for the external environment, are in many instances also applicable to the internal environment (which is, after all, mostly water).

(2) Temperature and life forms

a Thermal tolerance is the phenomenon of an organism's ability to withstand a certain temperature without being able to develop at this temperature. Obviously, this tolerance will be more extreme when the protoplasm

---

[1]Note, however, that epilimnion and hypolimnion are limnological terms that are not applied to the oceans.

contains a lower percentage of water, seeing as how the thermal properties of water are highly influential. The highest tolerance can be found in spores of bacteria (Dickson, 41), for example those of *Clostridium botulinum*, which retain their ability to develop and reproduce even after spending hours in a hot oil bath at 140 °C.

The lowest temperature that organisms can withstand is equally difficult to assess. Rahm (104) found in the Cryogenic Laboratory of the University of Leiden that organisms of a fairly high level of organization (such as rotifers, tardigrades, and nematodes) can withstand the extremely low temperatures of liquid helium (−271 °C) for 8 to 9 hours. Insect eggs were still 30% active after spending 3½ hours in liquid air (−190 °C). In the same laboratory, Zirpolo found that phosphorescent bacteria did not lose their viability after 10 hours at −271 °C, and even regained their phosphorescence when brought back to normal temperature. De Jong placed trypanosomes in liquid air (−190 °C) for weeks, and the organisms remained alive. Diatoms (Pictet), yeasts and fungi (Beyerinck), too, are resistant to extreme cold.

Blue-green algae, however, as resistant as they are to high temperatures and other extreme conditions, lose their blue-green pigmentation at these low temperatures.

It is known (Pictet, 100) that frogs and fish can remain frozen solid for long periods of time without dying, and in Arctic and Antarctic waters mussels, lobsters, water mites, and infusoria[2] can remain frozen for more than four months without adverse effects (Imhof, 68).

In nature, however, the temperature range is limited. It spans approximately 100 °C: from −40 °C during the Arctic winter to +60 °C on sunlit ground (even at our latitude) in summer. Many organisms have this limited tolerance, although the temperature range of various plants and animals has been measured much more precisely.

When one employs the terms "stenothermal" (having a narrow temperature range) or "eurythermal" (having a broad temperature range), it is important to note whether one is referring to latent or to vegetative life.

Vegetation (vegetable life) in temperatures above 100 °C and below 0 °C would only be possible in saline solutions (due to the higher boiling point and lower freezing point of salt water).

b It appears that *Penicillium* is indeed able to grow in cold saline waters below 0 °C. For extremely high temperatures we first look to hot springs, about which we often find conflicting information in the literature. (The older literature is summarized in the dissertation of Hugo de Vries, 121.) In 1929, when visiting Yellowstone National Park, van Niel and Thayer found[3] colorless threadlike bacteria (perhaps sulfur bacteria?) at 91 °C, while blue-green algae began to appear at 75 °C. Amoebae exist in springs in which

---

[2]Meaning small aquatic eukaryotes such as algae, ciliates, and small invertebrate animals like rotifers.
[3]Verbal report. [Original footnote from Baas Becking.]

the temperature is higher than 50 °C (8); diatoms, fly larvae, and purple bacteria also seem to be able to develop at these temperatures. The occurrence of fish and spiders in warm springs has also been described. One can thus categorize these organisms as being thermophiles.

Miehe recently repeated and expanded (85) his previous research (84) on hay heat,[4] in which he was able to isolate a large number of thermophilic organisms. The self-heating of organic materials that he studied in pure cultures produced surprising results. It appeared that the creation of heat through decomposition continued far enough to reach the temperature at which plant life ceased to exist.

The following table provides a representation of this.

Organism	Self heating up to:	Lethal temperature for organism
Rhizopus nigricans	38 °C	<40 °C
Penicillium glaucum	41	40–45
Aspergillus niger	53.5	50–55
Aspergillus fumigatus	57	>60
Mucor corymbifer	56.5	60
Thermoidium sulfureum	58	55–60
Actinomyces thermophilus	63	60–65
Thermomyces lanigunosus	62.5	60–65
Bacillus calfactor	74	>70

For a peculiar form of thermal tolerance by organisms in highly saline lakes, see Chapter X.

The number of thermophilic organisms is huge; bacteria worth mentioning include the bacterium which causes sulfate reduction (Elion, 49), as well as Actinomycetales and protozoa. Sulfate-reducing bacteria, blue-green algae, amoebae, and purple bacteria form an ecological group that appears to exist in another extreme environment, namely highly saline lakes (see Chapter X).

Most organisms, however, have a narrower temperature range. Organisms of the open oceans are generally stenothermal (i.e., bound to a narrow temperature range), whereas it is precisely the littoral beings found in coastal areas – exposed at low tide to more extreme conditions (adapted to a broad temperature range) – appear to be eurythermal. Organisms living in mudflats are also able to withstand large temperature fluctuations. Figure IV.3 shows some curious points on the "biological thermometer."

c *The vital functions within an organism's natural temperature range* are all temperature-dependent to some degree.

J. van Sachs was one of the first to look at the influence of temperature on different biological processes, coming to the conclusion that each process begins at a certain temperature (minimum), is most intense at a certain temperature (optimum), and ceases at a certain temperature (maximum).

---

[4]The heat created through the decomposition of hay.

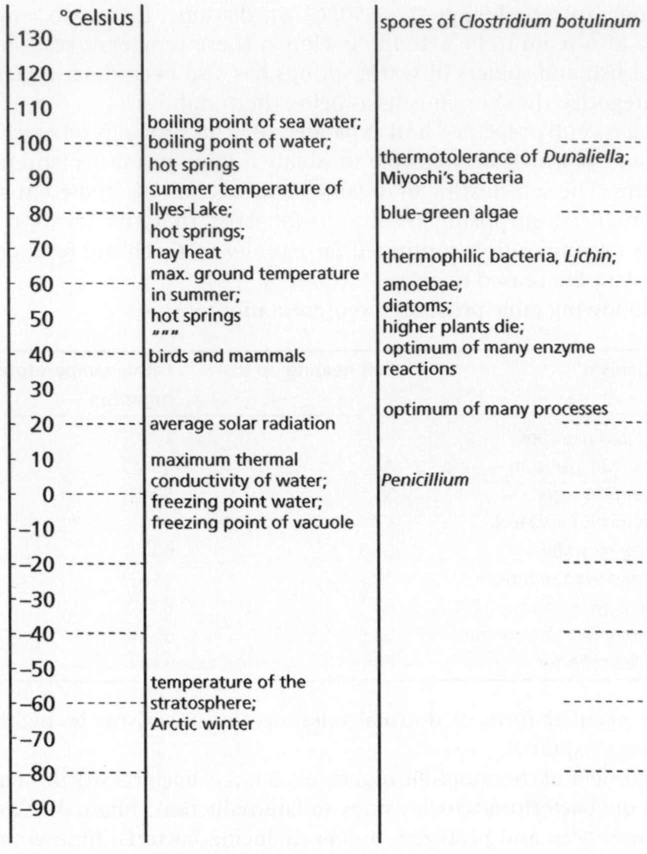

°Celsius		spores of *Clostridium botulinum*
130		
120		
110		
100	boiling point of sea water; boiling point of water;	
90	hot springs	thermotolerance of *Dunaliella*; Miyoshi's bacteria
80	summer temperature of Ilyes-Lake;	blue-green algae
70	hot springs; hay heat	thermophilic bacteria, *Lichin*;
60	max. ground temperature in summer;	amoebae;
50	hot springs """	diatoms; higher plants die;
40	birds and mammals	optimum of many enzyme reactions
30		
20	average solar radiation	optimum of many processes
10	maximum thermal conductivity of water;	
0	freezing point water;	*Penicillium*
−10	freezing point of vacuole	
−20		
−30		
−40		
−50		
−60	temperature of the stratosphere; Arctic winter	
−70		
−80		
−90		

Figure IV.3 Temperature ranges of a selection of organisms.

Van het Hoff[5] later gave a rule for the temperature dependence of the reaction time of chemical reactions. This reaction time should increase by 2 to 3 times for every 10 degrees of temperature increase. We will not discuss the derivation of this rule here, as it was well described by Arrhenius, but will simply point out that this "R–G–T rule" (*Reaktionsgeschwindigkeit-Temperatur*[6]) has also been applied to the intensity of the broadest range of biological processes.

Within certain limits, this rule seems to hold true for a large number of processes, yet at higher temperatures (above van Sachs' "optimum"), one sees

---

[5]Jacobus Henricus van het Hoff (or van't Hoff) (1852–1911) was a Dutch physical chemist and first winner of the Nobel Prize in Chemistry. Among other things, he worked on the temperature dependence of the rate of chemical reactions.
[6]Reaction Rate Temperature.

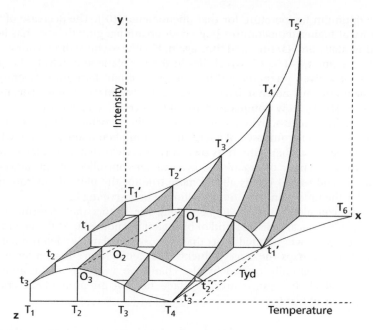

Figure IV.4 Link between intensity, time, and temperature.

a decrease in intensity [in the rate of the process] as temperature increases. Seeing as how organisms in the natural environment live in variable temperature conditions, van het Hoff's rule would be of little use to environmental science, were it not that – primarily through findings from research carried out in the Utrecht Botanical Laboratory – in a certain sense a link can be found between the R–G–T rule and the "optimum" phenomenon as established by van Sachs.

Figure IV.4 provides, in three dimensions, a schematic view of the link between the intensity of a biological phenomenon (y-axis), temperature (x-axis), and a third factor which shows the observation time (z-axis).

In the x–y plane, the curve of T_1'—T_2'—T_3'—T_4'—T_5' depicts the influence of temperature upon the intensity of a biological process according to van het Hoff's rule.

In this curve, the intensity continually increases as temperature increases. The curves of $t1$—0_1—t_1', t_2—0_2—t_2', and t_3—0_3—t_3' show the influence of temperature at the observation times t_1, t_2, and t_3. These have become "optimum curves," where the optima lie at 0_1, 0_2, and 0_3. The optimum is therefore dependent upon the duration of observation. A third group of lines is shown as the borders of the hatched planes (T_1'—T_1, T_2'—T_2, T_3'—T_3, etc.). They show how the intensity of a biological phenomenon changes over time at a given temperature. It is hereby assumed that for a given phenomenon, the temperatures T_1 and T_2 can be endured for an unlimited time, thus eventually making

T_2 the optimum temperature for that phenomenon (0_3). The decrease of this intensity at higher temperatures (e.g., with breathing, growth, etc.) has been experimentally proven time and time again. Finally, we must draw your attention to the line t_1'—t_2'—t_3', which lies in the y–z plane and depicts the time it takes for a given phenomenon to cease at a certain harmful temperature. This line is already known in bacteriology as the "sterilization line." The above discussion shows how complicated the influence of temperature is on just one biological process, when one considers the inevitable factor of time. Because life itself is composed of a large number of phenomena, each of which is influenced by temperature in its own way, it is clear that an explanation of the thermal behavior of organisms, even under controlled circumstances, is still far beyond reach. This is all the more true for the natural environment, where for the time being one remains limited in describing it.

Bayliss considers viscosity to be among the factors that determine the intensity of chemical reactions and of biological phenomena. At the beginning of this chapter we pointed out that viscosity is largely determined by temperature. Perhaps the huge influence that temperature has on biological phenomena is partially caused by the change in viscosity of the aqueous systems of which living structures are made; here we are primarily describing the properties of water.

EDITOR'S NOTES

In keeping with his geobiological theme, Baas Becking begins with a survey of how temperature influences the environments where organisms live. His considerations range from the influence of temperature on the viscosity of water, as this might affect the motility of small organisms, to the influence of seasonal variations in temperature on the stratification and mixing of natural water bodies.

Of particular interest are Baas Becking's reflections on the adaptation of organisms to temperature in the natural environment. He recognizes that various forms of life can survive temperature extremes ranging from boiling oil to liquid oxygen, and, even more interesting, he relates how organisms seem to thrive under a wide range of temperatures. He correctly surmises that organisms adapted to high-temperature environments are likely thermophiles, and summarizes how different types of organisms seem to have different maximum temperature adaptations, as revealed by studies of hot springs. He also discusses how new work in the field establishes that a typical organism increases its metabolic rate by a factor of 2–3 for a 10 °C change in temperature up to its temperature maximum, after which further increases cause sterilization. This is an accurate picture of how we still view the temperature response of organisms today.

Baas Becking seems dismayed, however, that many biological processes combine to yield the metabolic temperature response of an organism in nature, causing the thermal response of an organism to be "beyond reach." While we would agree today that Baas Becking was essentially correct – many factors contribute to the temperature response of an organism, including time – we are not so pessimistic. Indeed, subsequent work has shown that the temperature response of individual organisms in nature is well defined and reproducible. In fact, organisms are well adapted to their environment. Arctic regions, for example, are populated with so-called psychrophilic organisms (thinking of microbes) which can metabolize at $< 0\,°C$, with temperature optima of around $20\,°C$, and maximum temperatures of $30\,°C$ or so. At the other end of the temperature spectrum, so-called hyperthermophiles may lie dormant at temperatures below $60–70\,°C$ and have a temperature optimum of around $100\,°C$. Baas Becking would likely have been pleased to know that some of these microbes have been found to metabolize at temperatures of $115–120\,°C$. At higher temperatures, however, they become sterilized.

CHAPTER V

Environmental Factors: The Chemical Environment

(A) Gas

The gases which are of principal importance for the development of living systems are: oxygen, nitrogen, hydrogen, methane, and carbon monoxide, as well as carbon dioxide, hydrogen sulfide, and ammonia. In Chapter VI, which discusses the cycle of matter, we will encounter these gases as a metabolic product, and as matter that can be used either directly or by other means to build up a cell.

The above-mentioned gases differ in their level of solubility in water. O_2, N_2, H_2, CH_4, and CO are poorly soluble, while CO_2, H_2S, and NH_3 are highly soluble. Various other factors also influence the amount of a given gas in water. In addition to the level of solubility, one needs to consider biological factors. The formation of a gas and how it is used – i.e., the part this gas plays in the cycle of matter – are of greatest importance. Two physical factors, temperature and pressure, will be discussed. The solubility of a gas in water, in equilibrium with the atmosphere, is proportional to the partial pressure of that gas in the atmosphere (Henry's law). A slight decrease in barometric pressure is sufficient to make gas bubbles surface in our canals. However, this particular pressure phenomenon is not directly related to Henry's law.[1] More likely, it is the alleviation of local oversaturation. Furthermore, temperature is a large factor influencing solubility: when the temperature increases, the solubility of a gas decreases. At the boiling point of water solubility is extremely poor; one can "expel the gases" through boiling. An overview of the solubility of various gases at different temperatures can be found in the table below.

At higher temperatures, the gas exchange in water is remarkable in quantitative terms, which should be taken into account when considering the

---

[1]Henry's law relates the equilibrium concentration of a gas in solution to the partial pressure of the gas above the solution. It can be given as: $p = K_H C$, where p is the partial pressure of the gas, K_H is the Henry's law constant at a given temperature, and C is concentration.

Baas Becking's: Geobiology, Or Introduction to Environmental Science, First Edition. Edited by Don E. Canfield.
© 2016 John Wiley & Sons, Ltd. Published 2016 by John Wiley & Sons, Ltd.

biological communities living in hot springs. Without engaging too much in further discussion of the other factors which influence the solubility of gases, it must be pointed out that soluble salts generally decrease the solubility of gases. This is especially visible with concentrated brines, which, in this case, can be compared to heat sources [in how the salt influences gas solubility] (see Chapter X, *Brine*).

Gas volumes under pressure of 1 atmosphere at 0 °C, soluble in a given volume of water. (Bunsen & Carius)

Temp.	0°	5°	10°	15°	20°
Oxygen	0.04114	0.03717	0.03250	0.02989	0.02838
Nitrogen	0.02035	0.01838	0.01607	0.01478	0.01403
Hydrogen	0.01930	0.01930	0.01930	0.01930	0.01930
Methane	0.05449	0.04993	0.04372	0.03909	0.03499
Carbon monoxide	0.03287	0.02987	0.02635	0.02432	0.02312
Carbon dioxide	1.7987	1.5126	1.1847	1.0020	0.9014
Hydrogen sulfide	4.3706	4.0442	3.5858	3.2326	2.9053
Ammonia	1049.6	941.9	812.8	727.2	654.0

Especially in the field of environmental sciences, one needs to realize that solubility, in terms of liquids in equilibrium with gases, as determined by physicists and chemists, is only a point of departure. In nature, both over- and under-saturation occur more often than not. Bodies of water can contain three times more oxygen than they would in a state of equilibrium.

The absorption and emission of gases would be an extremely slow process if it were to occur only through diffusion. Gases with poor solubility are absorbed at a very slow rate, while highly soluble gases are emitted slowly. For example, to replace the carbon dioxide dissolved in water with hydrogen, this water would need to be vigorously shaken with hydrogen for at least 24 hours.[2] It is therefore understandable that the state of a gas in water (be it a raised-bog lake, city canal, shallow and stagnant fresh or salt water, or the hypolimnion [of a lake], see Chapter VIII) has little or nothing to do with its gaseous state in the atmosphere. However, we still believe that free-flowing water (such as the ocean,[3] a fast-flowing stream, or a lake rippled by the wind) is generally in equilibrium with the atmosphere. This is of course only true where the influence of living organisms is not the dominating factor. Thus, where biological and geochemical influences are of secondary importance, water is in equilibrium with an atmosphere of a remarkably constant composition, namely:

---

[2]This is in a part because carbon dioxide in water is in equilibrium with charged bicarbonate and carbonate ions, as discussed later in this chapter. To understand the efficiency of CO_2 emission from water, one must also consider the kinetics of equilibrium between the various forms of inorganic carbon species in water.

[3]Baas Becking is probably referring to the upper ocean.

20.94 volume percentage oxygen
0.03 volume percentage carbon dioxide[4]
79.03 volume percentage nitrogen
(0 °C, 760 mm)

However, owing to its specific solubility, air is of a different composition when it is dissolved in water. If this air, obtained by deaeration, is analyzed, then instead of 21% O_2 it will consist of approximately 35% O_2; the level of carbon dioxide can also show a dramatic increase. Thus, water in equilibrium with air is relatively rich in oxygen and carbon dioxide, but poorer in nitrogen than is the air.

Regarding its gas exchange, a given body of water can, as a unit, be seen almost as an organism in which cycles take place. It does not in fact matter which gas one chooses as a starting point to examine these processes. Within the scope of this discourse, however, we will limit ourselves to an easily soluble gas, carbon dioxide, and a poorly soluble gas, oxygen.

Gases such as carbon dioxide, hydrogen sulfide, and ammonia are very soluble in water as they make a chemical bond with the water. It is commonly assumed that CO_2 reacts with water:

$$CO_2 + H_2O = H_2CO_3$$

and that this compound partially dissociates:

$$k_1[H_2CO_3] = [H^+][HCO_3^-] \tag{V.1}$$

in which k_1, the first dissociation constant of carbon dioxide at 20 °C, is approximately 3.5×10^{-7}. With this first dissociation, bicarbonate is created, which further dissociates to carbonate:

$$k_2[HCO_3^-] = [H^+][CO_3^{2-}] \tag{V.2}$$

in which k_2 the second dissociation constant, amounts to 4.7×10^{-11}.

If we assume that generally all of the CO_2 has become H_2CO_3, then in a natural body of water, carbon dioxide is present in three different states, namely undissociated, as bicarbonate, and as carbonate. In both equations (V.1) and (V.2) the concentration of hydrogen ions is also visible, and thus we can decide that the hydrogen ion concentration (pH) will have a large influence on the ratios of the three states in which carbonic acid occurs. These ratios are easy to determine using a method introduced by Michaelis.

When we set the total amount of carbonic acid as equal to 100, it can be deduced that:

$$[H_2CO_3] + [HCO_3^-] + [CO_3^{2-}] = 100 \tag{V.3}$$

---

[4]Now 0.04 volume percent.

And (3) combined with the equations (V.1) and (V.2) gives:

$$\left[H_2CO_3\right] = \frac{100}{1 + \dfrac{k_1}{\left[H^+\right]} + \dfrac{k_1 k_2}{\left[H^+\right]^2}} \tag{V.4}$$

$$\left[HCO_3^-\right] = \frac{100}{1 + \dfrac{\left[H^+\right]}{k_1} + \dfrac{k_2}{\left[H^+\right]}} \tag{V.5}$$

$$\left[CO_3^{2-}\right] = \frac{100}{1 + \dfrac{\left[H^+\right]}{k_2} + \dfrac{\left[H^+\right]^2}{k_1 k_2}} \tag{V.6}$$

The equations (V.4), (V.5), and (V.6) make it possible to determine the ratio of carbonic acid, bicarbonate, and carbonate at a given level of acidity (see table below and Figure V.1), using the pH notation.

Molar percentage of carbonic acid, bicarbonate, and carbonate at different levels of acidity

$[H^+]$	10^{-4}	10^{-5}	10^{-6}	10^{-7}	10^{-8}	10^{-9}	10^{-10}	10^{-11}	10^{-12}
H_2CO_3	99.6	96.6	74.1	22.2	2.8	0.3	—	—	—
HCO_3^-	0.4	3.4	25.9	77.8	96.7	95.2	67.9	17.6	2.1
CO_3^{2-}	—	—	—	—	0.5	4.5	32.1	82.4	97.9

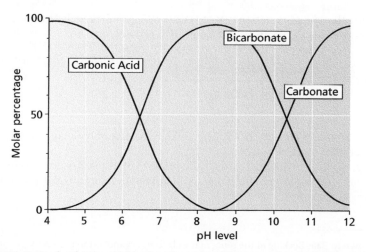

Figure V.1 Ratio of carbonic acid, bicarbonate, and carbonate as a function of the various acidity levels.

From the figure it is possible to read some facts important for our use. First, the free carbon dioxide is practically nonexistent when pH > 9. Second, bicarbonate is at a maximum level when pH = 8.2.

Third, when pH = 12, practically all the inorganic carbon is present as carbonate.

Seeing as how the green organisms and autotrophic bacteria absorb carbon dioxide from water, from the figure it can be deduced in what state it is available to them,[5] depending on the pH value. The fact that a large proportion of green plants are no longer able to assimilate at pH levels > 9 means that the state in which the carbon dioxide is available is very important for these life processes.

If the total amount of CO_2 contained in one liter of water and the pH level are known, this information can be used to determine the percentages of the various states of CO_2 in the water.

In order to delve deeper into the mechanisms of the carbon dioxide gas exchange, we must make use of the term "base excess."[6] By this term, we understand Johnston (71) to mean the number of equivalent metal ions (B^+) in a certain amount of water that is in equilibrium with the hydrogen, hydroxyl, bicarbonate, and carbonate ions, thus:

$$[B^+]+[H^+] = [HCO_3^-] + \frac{[CO_3^{2-}]}{2} + [OH^-] \tag{V.7}$$

The ion product for water (K_w) is constant (approximately 10^{-14} at room temperature), or

$$[H^+][OH^-] = K_w \tag{V.8}$$

With the aid of equations (V.1), (V.2), (V.7), and (V.8) it is now easy to express the base excess in [H_2CO_3] and [H^+], arriving at:

$$[H_2CO_3] = \frac{\{[B^+]+[H^+]\}[H^+]^2 - K_w[H^+]}{k_1[H^+] + \frac{1}{2}k_1k_2} \tag{V.9}$$

This equation (Johnston, 1916) enables us to determine many phenomena of water with more accuracy.

First of all, the equation predicts the pH level of distilled water or rainwater in equilibrium with the atmosphere. At room temperature, [H_2CO_3] in distilled water is approximately 10^{-5} molar.

---

[5]Unknown to Baas Becking at the time, many algae and cyanobacteria can actively concentrate bicarbonate into the cell, converting it further to CO_2 for use by the organism.
[6]This is the same as alkalinity.

Seeing that [B$^+$] is very small in distilled water or rainwater, equation (V.9) can be simplified to:

$$10^{-5} = \frac{[H^+]^3 - K_w[H^+]}{k_1[H^+] + \frac{1}{2}k_1k_2}$$

or, when we enter the values of k_1k_2 and K_w:

$$10^{-5} = \frac{[H^+]^3 - 10^{-14}[H^+]}{3.5 \times 10^{-7}[H^+] + 0.81 \times 0^{-7}}$$

If we neglect the terms 10^{-14} [H$^+$] and 3.5×10^{-7} in this fraction, we arrive at:

$$3.5 \times 10^{-12} = [H^+]^2,$$

Or

$$[H^+] = 1.87 \times 10^{-6}$$

The negative logarithm of this amounts to $-(0.272 - 6) = 5.278$. Rainwater or distilled water, in equilibrium with the atmosphere, would have a pH of around 5.7, something that is easily proved in reality.

The base excess, the so-called "alkalinity" of the water, can be determined by titrating the water with acid until the bicarbonate has practically disappeared, thus amounting to a pH of 4.5 (see Chapter IX). The numerator of the fraction in equation (V.9) will only be zero for very small values of [H$^+$], meaning that even in very alkaline waters traces of H$_2$CO$_3$ can still be found. Thus if free H$_2$CO$_3$ is needed for carbon fixation, this will be available, albeit in very small quantities. However, if H$_2$CO$_3$ is absorbed during the carbon fixation at a constant of B$^+$, it follows from equation (V.9) that the acidity will be reduced, while from Figure V.1 it is clear that with decreasing acidity the equilibrium between carbonic acid : bicarbonate : carbonate will move towards the carbonate. When the carbonic acid is removed from the water, the bicarbonate will dissociate further: $2HCO_3^- \rightarrow H_2CO_3 + CO_3^{2-}$, but seeing that now the total amount of carbon dioxide has been reduced, the ratio between the three states of carbon dioxide has been altered.

The other way around, when respiration has the upper hand, the amount of carbon dioxide in the water will increase. A part of the respired carbon dioxide will dissociate to bicarbonate and even to carbonate, but the increase of the total amount will result in a shifting of the equilibrium towards the carbonic acid, which, in accordance with equation (V.9), will be accompanied by an increase in the acidity (lowering of the pH) of the water.

Figure V.2 displays the influence of carbon assimilation and respiration on the acidity levels of water. A green alga (*Ulva lactuca*) with a surface area of 100 cm^2 was placed in an aquarium, with 1 liter of seawater at a temperature of 20 °C. The pH of the water in light (and in a different experiment in the dark)

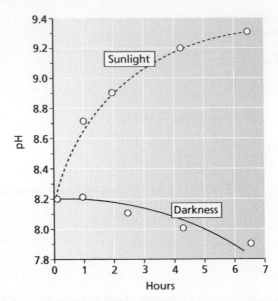

Figure V.2 Influence of light and darkness on the change in acidity levels of seawater by the alga *Ulva*.

was measured as a function of time. It was observed that the change that takes place in the presence of sunlight (the increase in alkalinity due to carbon dioxide extraction) is a much faster process than that which takes place in darkness (acidification due to carbon dioxide production). This phenomenon is caused by the buffering capacity of water and includes the ability of certain solutions (mixtures containing bicarbonate, carbonate, or hydrogen phosphate and dihydrogen phosphate) to absorb hydrogen or hydroxide ions without changing their acidity levels (pH). The role that the carbonic acid equilibrium plays in the precipitation of limestone will be further discussed in the chapter on seawater.

While the water still contains carbon dioxide, the amount of oxygen in it can reach low levels, close to zero. This approximates the situation in black mud, which consists partially of colloidal ferric sulfide, and where free oxygen does not occur. On the tidal mudflats, when worms burrow through the mud, the walls of their tunnels are often covered with reddish ferric hydroxide which comes from the oxidation of iron-rich mud. The single-celled green alga *Trachelomonas*, when placed in water containing black mud, will, in the light, cover itself with a layer of ferric hydroxide which is created through the expelled oxygen. Also, in places where large amounts of organic material are decomposing, oxygen can almost completely disappear. On the other hand, there are also many waters (especially stagnant warm pools containing large quantities of algae) in which the oxygen excess can reach over 200%. Stirring the water with a stick is enough to dissipate the bubbling gas. The organisms which live very close to the surface on the tidal mudflats experience extreme fluctuations in oxygen levels. When the tide ebbs, the absorption of air is even audible as it replaces the receding waters that had filled the capillaries of the sand or clay. The remaining pools are soon oversaturated with oxygen due to the activities of green plants. The active carbon dioxide assimilation is also responsible for the increase in pH. With the incoming tide, the soil capillaries once again fill with water, and the black mud (a product of bacterial sulfate reduction) displaces the oxygen, which is only partially replenished by the incoming water. This explains the extreme fluctuations in oxygen levels in the upper layers of the sediment.

(B) Water

At the beginning of this chapter we discussed the dissociation of water and the hydrogen and hydroxyl ions that occur as a result. This brief discussion is not sufficient, however, to fully understand water. Indeed, water is apparently not yet chemically characterized. In the series of compounds with low molecular mass, water holds a special position. For a molecular weight of 18, its boiling and melting points are too high. These abnormal traits have induced certain scientists (e.g., Armstrong, 3) to view water as a mixture of three polymers, namely H_2O (monohydrol), $(H_2O)_2$ (dihydrol), and $(H_2O)_3$ (trihydrol). Various modern studies seem to confirm the complicated structure of the water molecule and also provide a reason for the abnormal freezing and boiling points, as well as for its maximum density being at $3.96\,°C$.[7]

At higher temperatures, water would contain simpler molecules, and at lower temperatures its molecules would be more complex. According to some researchers, newly melted water would contain the more complex water molecules for quite some time still, while newly condensed steam holds the characteristics of the monohydrol for many hours (Barnes, 17). Lloyd and Barnes (18) claim that the suitability of water for living organisms (protozoa and green algae) depends on the amount of its complex molecules, which seem to be very suitable for sustaining life, while the monohydrol has a toxic effect. The high levels of dihydrol in the water would also explain the blooms of plankton at the edges of melting ice. Lloyd and Barnes' ideas are still largely speculative, but it remains a possibility that the complex nature of water exerts an influence on the organisms living within it. New data has shown that the recently discovered hydrogen isotope [deuterium, ^2H] creates a very different type of water, called "heavy water." The freezing point and melting point of this water

---

[7]The abnormal properties of water are due to the structure of the water molecule, which produces an asymmetry in change, inducing a weak hydrogen bonding between water molecules. This was described for water in 1920 by Latimer and Rodebush (Latimer, Wendell M., & Rodebush, Worth H. (1920). *Journal of the American Chemical Society* 42 (7): 1419–1433. doi:10.1021/ja01452a015), but this new insight had apparently not filtered into the environmental sciences at the time of Baas Becking's writing.

are higher (freezes at 3.8 °C), while it seems to be toxic for higher organisms (109).[8] We now see water as not merely H_2O but as a connection between two types of hydrogen as well as a mixture of polymers, including ions. Furthermore, when water is in contact with the atmosphere, the gases nitrogen, oxygen, and carbon dioxide can be regularly found in the water.

If other substances are dissolved in water, this water will also have its effect on these materials. For instance, the dissociation and the osmotic values of these materials are affected as much by the nature of the materials as by the nature of the water. Before posing the question of which materials should be regarded as essential for life, the role of water as water vapor, of dissociation, and of osmotic values should be further reviewed.

(C) Water vapor

Many organisms are able to absorb water in a gaseous or cloud state. The *Sequoia sempervirens* (giant redwood) occurs in California's coastal zone in so-called tule fogs, in valleys in which the moist sea winds carry in large amounts of water vapor. Plants that live on trees or rocks, such as the green algae *Klebsormidiales* and *Trentepohlia*, and also lichens, are totally dependent for long periods of time on moisture taken from the atmosphere. An atmosphere totally saturated with moisture is still not able, however, to replenish the cells with sufficient water. For this, contact with liquid water is necessary. Peat moss (*Sphagnum*) also has the ability to absorb water vapor. The amount of water vapor absorbed per unit of time per surface-area unit depends on the level of saturation of the peat moss and the humidity level of the atmosphere. Even if the atmosphere is completely saturated with water vapor, however, the plant is still unable to saturate itself completely with water, as can be seen in the table below.

Living *Sphagnum*, dried above strong sulfuric acid, was placed above saline solutions with a known vapor pressure at ± 20 °C. The weight became constant after some time:

Above	water	Vapor pressure	100%	3	× Dry weight
"	1 mol NaCl	" "	96.4%	2	× " "
"	2 mol NaCl	" "	91.3%	1.50	× " "
"	3 mol NaCl	" "	86.6%	1.33	× " "
"	4 mol NaCl	" "	79.5%	1.25	× " "
"	5 mol NaCl	" "	69.2%	1.20	× " "

The dry plant, after submersion in water, gained 2× its original weight (experiments conducted in the Botanical Lab in Leiden by Ms. A. Krythe and H. D. Verdam).

---

[8] This is due, in part, to the influence of heavy water on the kinetics of enzymatic reactions. Heavy water also interferes with mitosis in eukaryotic cells, inhibiting reproduction.

When the vapor pressure of a saline solution[9] is equal to the atmospheric water vapor pressure, the influences of these two different environments on organisms will differ. It is the contact with the liquid water molecules that seems to have a special effect. In general it can be said that solutions with a vapor pressure of less than 85% are highly toxic for most organisms (124). The brine organisms are an exception to this rule (see Chapter X).

In some cases water is apparently not absorbed either as liquid or as vapor by these organisms. With the larvae of the common clothes moth, the greater wax moth, and possibly also the larvae of the petroleum fly (Chapter X), the water present within these organisms originates from the metabolization of organic matter (metabolic water).

(D) pH level

The acidity level of a body of water is determined by the concentration of H^+ ions and is expressed as the pH level. The pH of natural water bodies and of soil extracts is very different. Figure V.3 shows a table with an overview of the pH levels occurring in various natural environments.

With the draining of the Wieringermeer polder, a layer of clay containing sulfur was found which, after oxidation in air, released so much sulfuric acid that solutions of over 0.1 N H^+ occurred (pH 0–1). The oxidation of this sulfur is caused by *Thiobacillus thiooxidans*, a bacterium first described by Waksman and isolated from the Wieringermeer clay by K. Griffioen and Ms. T. Hof. The low pH level (0–2) of some soils must be ascribed to the activity of this bacterium, except when acid waters are formed by volcanic exhalation (HCl, SO_3).

Raised bog, acidic humus, and heath have a pH value of 3–5, the cause of which is not entirely known (see Chapter VII). During or shortly after a thunderstorm, rainwater can have a high acidity level (nitric acid?). As mentioned earlier in this chapter, rainwater usually has a pH of around 5.7. Blood (pH 7.6) is close to being neutral, while seawater is a solution with a remarkably constant pH level (pH 8.1–8.2). Among the more alkaline solutions, the sodium lakes are worth mentioning, where the alkalinity is caused by dissolved soda or trona (Na_2CO_3, $NaHCO_3$).

Figure V.3 displays the range wherein various substances occur in solution, depending on the pH level. The solubility of aluminum salts is very low when pH > 3, for ferric salts when pH > 4, and for ferrous salts when pH > 6.5. Calcium carbonate, magnesium hydroxide and magnesium carbonate precipitate at much higher pH levels. However, it must be remembered that traces of these salts, even with low acidity levels [high pH], are still present in solution. For instance, the existence of iron bacteria at a pH level of 9.2 has been proven (Meehan, 83), while at this pH the ferrous ion is only marginally present.[10]

---

[9]Baas Becking is referring here to water activity, which decreases as salinity increases.
[10]In fact, the oxidation rate of ferrous iron increases with increasing pH, and one would expect only very short life times, and exceeding low concentrations, for ferrous Fe in solution at high pH in the presence of oxygen.

pH	Natural Occurrence	Chemical Data	Algae	Higher Organisms
12				
11	Lake, Egypt (Braun-Blanquel)			
10	Lake, Marina, Calif (B. Becking)	$Mg\,CO_3$ prec.		Halophytes Desert plants
9	Lake, Searles, Calif (B. Becking)	$CaCO_3$ prec.	*Dunaliella*	*Populus*
8	Sea water		*Ulva lactuca Enteromorpha*	Barley
7	Blood, neutral point	Carbonate limit	*Cladophora Spirogyra*	Sugar beet *Cirsium anglicum*
6		Ferro-limit		Peas
5	Rain, distilled water	Bicarbonate limit		Red clover Oat
4	Humus	Ferri-limit	*Odogonium Itziagsohni*	Buckwheat *Molinia*
3	Pine heath Thunderstorm rain	Al^{3+}limit		Andromeda *Vaccinium myrtillus*
2	Soil, Goldcoast (F.J. Martin) Sand, Denmark (O. Jansen)			
1	Volcanic lakes			
0	Oxidized marine clay *Wieringermeer (Marmsen)*			

Vertical ranges in "Natural Occurrence": field- / forest- / lake-; Natron; Sea water; Fresh water; Deciduous; Agricultural; Raised bog; Temperate coniferous forest.

Vertical labels in "Chemical Data": no Ca^{++} and Mg^{++}; much CO_3^{--} and OH^{++}; $HCO_3^- - CO_3^{=}$; Ferro; CO_2; Ferri.

Vertical labels in "Algae": Cyanobacteria; Desmidieaceae.

Vertical labels in "Higher Organisms": *Gasterosteus*; Fish.

Figure V.3 pH as an environmental factor.

The chemical environment is influenced in large part by the actual[11] pH of the solution. The actual solution is furthermore an important factor for the existence of living organisms. For a number of processes, including lactic acid fermentation, it has been shown that it is not the concentration of hydrogen ions but the amount of titratable acid that limits the growth of organisms. However, the actual pH remains of great importance for a great number of living organisms.

The acid-loving peat moss (Spagnaceae) can be classified according to its occurrence at various pH levels. This is how we know that certain algae and also certain higher plants find a limit to their growth, occurring only within a certain pH range or only on one side of the balance, be it acidic or alkaline. Several examples are given in the last column of Figure V.3.

It is odd that organisms which have adapted to other extreme environments (hot springs, salt lakes) also display their hardiness with regard to a variation in the pH (e.g., amoebae, blue-green algae, seaweed, flagellates). The stickle-back (*Gasterosteus*) is another example of an organism capable of living in an

---

[11]The original Dutch is "*actueele.*" Baas Becking seems to be referring to the buffering capacity of a solution.

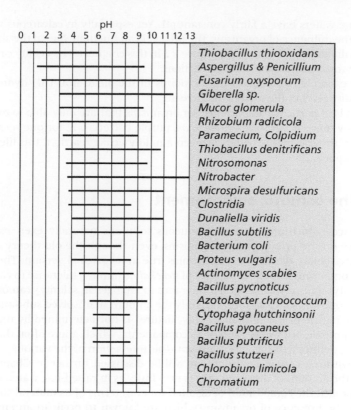

Figure V.4 pH as an environmental factor.

environment which has great variations [in its chemical characteristics]. This fish has been found both in very saline waters (7%) and in acidic peat water.

Figure V.4 shows the pH range for a number of microorganisms, the data for which have been derived mainly from the journals of Waksman (122) and Russel (108), as well as from the work of C. B. van Niel (91) on colored bacteria, and from the author on the halophile alga *Dunaliella* (13).

Various organisms in this table occur in close interaction with higher plants (symbiosis): *Bacillus radicicola*[12] with the legume family, *Phycomyces* (algal fungi) with heath, orchids, etc., Actinobacteria (ray fungus) with alder and sea-buckthorn, higher fungi (Basidiomycota) with many of our trees, etc.

The distribution of the above-mentioned higher plants is partially dependent on the ability of their symbiotic partners to thrive. This needs to be kept in mind if a study on the distribution of plants in symbiosis with microbes is undertaken.

---

[12]Now placed in the genus *Rhizobium*.

Some waters have a fairly constant pH. Yet especially in calcareous water, under the influence of organisms, the pH level can fluctuate due to the shifting carbon dioxide equilibrium (see Figure V.2 in this chapter). The acidity can also change through purely chemical reactions.

For example, when black mud oxidizes in the air to iron (III) hydroxide, the acidity level of the water will increase.

This is of particular importance for organisms that are only able to exist in a narrow range of pH (e.g., *Chromatium*, purple bacteria). If, for one reason or another, the solution experiences even a mildly acidic reaction, this life form will disappear.

(E) The osmotic environment

Substances which are present in aqueous solutions naturally exert osmotic pressure on the protoplasm of organisms, even though the old theory of the extractive ability of salts no longer holds true in its original version. The division of organisms into the categories of *stenohaline* (limited tolerance to changes in salinity) and *euryhaline* (broad tolerance to changes in salinity) can only be made if one takes into account not only the amount of dissolved substance but also the nature of the substance. This "nature" is not determined by the total amount of salt, but by the ratio of its components, which, more often than the amount, defines a certain environment. This interaction of the various components within a certain environment will be discussed further in Chapter IX. When the amount of dissolved salts is very little, as in raised-bog water, where we approach the composition of distilled water in terms of its mineral content, we see the presence of organisms which are known to occur in an environment with much higher salt concentrations. Flagellates which generally occur only in brine (*Dunaliella*) have also been spotted in raised-bog water. Also, the marine crab *Evadne* has the ability to stay alive for a long period of time in distilled water.

With terrestrial plants, the tissues are not saturated with water, but the "suction force" of the cells has a certain value, which increases upon wilting and approaches zero upon reaching water saturation. This suction force is caused in part by the osmotic value of the cell fluid. It is, however, also possible, and even likely in many cases, that the colloids or the cytoplasm itself can bind water. This "bound water" will become inactivated on a plasma surface, and therefore the measured osmotic value is caused not only by the dissolved matter but also by the state in which the water is found within the cell.

The so-called water permeability of the protoplasm, i.e., the speed with which water enters or leaves the protoplasm, is yet another factor that needs to be considered in environmental science. It is possible that the rate of dehydration for aquatic plants (e.g., seaweed, which is exposed at low tide) is also determined by this factor.

The phenomenon of osmosis itself, in the absence of the above-mentioned secondary factors, is rarely visible. Yet the terms "water absorption with a high inner [solute] concentration" and "water release with high outer [solute] concentration" has a high heuristic value; the mechanism for keeping the cell fluids as well as the bodily fluid constant appears to be based on highly complex processes. The eel, for instance, has a specific mechanism in its gills to extract the NaCl out of the blood. However, the functions of the kidney, the pulsing vacuole of protozoa and lower plants, and other excretory organs are not yet well known.

If we measure the osmotic environment in atmospheres, we get:

For blood: 6.8 atm.
Seawater: 28.0 atm.
Brine 200: 300 atm.,

while river water has an osmotic value of only 0.2–0.3 atm. The osmotic value of the cell fluid (cytoplasm) is variable in terrestrial plants. Values of 6–8 atmospheres have very often been found. These values are much higher for halophytes and in specialized cells (e.g., stomata in plant leaves), and the cytoplasm can also consist of a more concentrated solution.

(F) The elements and connections necessary for life

The factors mentioned in the title of this section are actually what form the chemical environment in a narrow sense. In Figure V.5 the black shaded line encircles the elements which have been proven to be crucial for certain organisms. The shaded squares depict the elements which are practically always found within living organisms. Both boundaries are to the best of our current understanding and should not be seen as definite. The number of known elements – some of them essential for life – continues to increase as our knowledge of their functioning grows and our analytical methods improve. Furthermore, between the "ubiquitous" elements and the elements which are "only necessary in special cases" there are subtle connections, such that a particular element can be of primary importance for one large group of organisms while being completely absent within others. Regarding the special function of most of the elements, much is still unknown, though for many a specific role has been proven. As can be seen in Figure V.5, most "vital" elements have a small atom, and therefore a low atomic weight. Indeed, the vital elements are rare above atomic number 20 (calcium).[13]

After Mendeleev had shown that the elements with low atomic weight were also the ones that occur most frequently on Earth, Preyer, and especially Sestini, proved that the elements with low atomic weight also occurred most frequently in living organisms. Errera summarized the older literature on this topic in 1910 in a discourse (51) entitled: *"Pourquoi les elements de la matière vivante ont-ils des poids atomiques peu élevés"*?[14] It is difficult to discern if the

---

[13]This is now known not to true, as many transition metals such as Mo, Zn, Mn, Cu, V, and Fe have a critical role in constructing biological enzymes.

[14]Why do the elements that occur in living organisms have a low atomic weight?

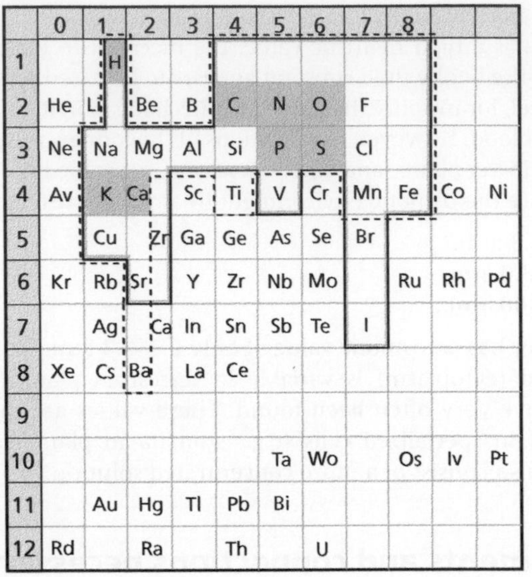

	0	1	2	3	4	5	6	7	8		
1		H									
2	He	Li	Be	B	C	N	O	F			
3	Ne	Na	Mg	Al	Si	P	S	Cl			
4	Av	K	Ca	Sc	Ti	V	Cr	Mn	Fe	Co	Ni
5		Cu	Zn	Ga	Ge	As	Se	Br			
6	Kr	Rb	Sr	Y	Zr	Nb	Mo		Ru	Rh	Pd
7		Ag	Ca	In	Sn	Sb	Te	I			
8	Xe	Cs	Ba	La	Ce						
9											
10						Ta	Wo		Os	Iv	Pt
11		Au	Hg	Tl	Pb	Bi					
12	Rd		Ra		Th		U				

Figure V.5 The "vital" elements and most frequently occurring elements.

suitability to serve as a biogenic substance coincides with a low atomic number or if life uses the most commonly occurring elements (in Figure V.5 marked off by a dotted line); "making do with what's on offer," so to speak.

The table below, assembled in accordance with data from Clarke and Washington (37), provides a list of the occurrence of the elements in the Earth's crust and in living systems.

It is possible to view the role of the various elements in the chemical environment from different points of view: first, as sources for the synthesis of organic bonds, among which C, H, O, N, S, and P are notable; second, as materials (often ions) which influence traits of the protoplasm (e.g., permeability). Essential elements can also negate each other's toxic influence (antagonism), act as catalysts in metabolic processes, and instigate reactions within the cell or elsewhere in the organism (hormonal). Furthermore, various substances can accumulate in an organism, more or less accidentally, as various analyses of similar organs and species have shown.

Element	Average % of lithosphere, hydrosphere, and atmosphere (from Clarke)	Composition of man (from Lotka)	Composition of higher plants (average from various analyses)
O	50.02	63.03	72.17
Si	25.80	trace	0.40
Al	7.30	trace	0.05
Fe	4.18	0.01	0.30
Ca	3.22	2.50	0.57

Element	Average % of lithosphere, hydrosphere, and atmosphere (from Clarke)	Composition of man (from Lotka)	Composition of higher plants (average from various analyses)
Na	2.36	0.10	0.13
K	2.28	0.11	0.75
Mg	2.08	0.7	0.20
H	0.95	9.90	9.33
Ti	0.43	—	—
Cl	0.20	0.16	0.30
C	0.18	20.20	14.33
P	0.11	1.14	0.27
S	0.11	0.14	0.25
F	0.10	0.14	—
Ba	0.08	—	—
Mn	0.08	—	—
N	0.03	2.50	0.80
Sr	0.02	—	—
All others	0.40	—	—

The chemical environment can also be viewed quantitatively, whereby, in addition to osmosis and antagonism, the minimum factors must also be taken into account. As early as 1840 J. von Liebig pointed out that when using fertilizer, the harvest increases only when the mineral dose is added within a certain range (77), and that larger applications [of fertilizer] have no further impact on the harvest. Liebig speaks of a factor "within the minimum" (das Gesetz des Minimums).[15] Blackman (30) calls such a factor "limiting" and establishes the law that "a life process influenced by a large number of factors, evolves in proportion to the factor within the minimum." While Liebig, in his book Agrikulturchemie, was looking for examples, Blackman proved the law for carbon dioxide assimilation. In the natural, chemical environment, substances are regularly found whose appearance is variable and whose concentration is low, which is the reason why these substances often are the limiting factor for various expressions of life.

For instance, in seawater the phosphate ion and the nitrate ion often are the limiting factors (see Chapter IX). On mineral-poor soil (peat colonies), copper can act as the limiting factor, while on alkaline soils it can be iron deficiency (78).

Sometimes a single substance, or a couple of substances, can predominate in the environment (NaCl, FeS, and petroleum), while in other cases (river water, normal arable soil) there is no substance that predominates. A more elaborate overview of the chemical environmental factors can be found in the works of Tschermak (119) or Kostychev (76). But here we will satisfy

---

[15]The law of the minimum.

ourselves with a description in a table of various biogenic elements with some of their specific attributes.

From this table it follows that the role of most biogenic elements is multiple, but that the substances which are of "secondary importance" are still the specific limiting environmental factors for large groups of organisms. There are some bacteria that seem to have the least [stringent] chemical needs/requirements, of which independence from calcium has been proven, and the independence from sulfur seems likely. Of the primary important elements only seven remain, namely C, O, H, N, K, Mg, and P.[16]

Elements which also hold a special place are those which Vernadsky calls "*éléments concentrateurs*"[17] namely, those which only occur as traces in the natural environment, yet in the inner environment [of the cell or organism] are often found in concentrations tens of thousands of times greater. Iodine, bromine, and copper are some of these elements.

Both the stimulating and toxic effects of a number of substances and elements in very low concentrations (oligodynamic effect) has been proven (e.g., for Zn, As, Ag, Cu, and Mo).

It must also be pointed out that modern analytic methods have been able to show the omnipresence of various elements and that each of these elements might, in a particular chemical form, be able to exert influence upon a given biological process. This makes it more complicated to obtain an overview of the meaning of the chemical environment and further removes us from the "old-fashioned" notion which attributes a specific function to each element. It is likely that at certain critical moments in the life of an organism trace amounts of certain elements must be crucial. In this light, the comparison of biogenic elements with the most common earthly elements seems to be but a very crude approach.

As an example of a chemical environmental factor which determines a very specific community, we will discuss black mud, which occurs in fresh water, seawater, and brine as well as in heat sources.[18] The sulfates which are reduced by heterotrophic bacteria (i.e., in the presence of organic matter) in an oxygen-free environment deliver sulfide, which, with iron salts, forms the hydrated ferrous sulfide (hydrotroilite). This material, mixed with sand or clay, occurs where there is an abundance of rotting organic matter (e.g., sea grass) and has the curious characteristic that it not only has a strong oxygen absorbing capacity, but it also expels hydrogen sulfide. With the oxidation of mud the ferrous oxide becomes iron (III) oxide. The energy released with this reaction is clearly utilized by a group of autotrophic bacteria (iron bacteria).

When "active" black mud is placed on a dish, the black color will disappear quickly due to oxidation. If part of the mud is covered with a

---

[16]In modern thinking sulfur would certainly be added to this list.
[17]Concentrator elements.
[18]Presumably Baas Becking is referring to hydrothermal areas.

Element	Occurrence in the external environment	Occurrence in the internal environment	Importance in the external environment	Importance in the internal environment	Location in the cell
N	N_2, NH_4^+, NO_3^-, or organic	Protein, alkaloid, cyanide, mustard oil, chlorophyll, etc.	As NO_3^-, often minimum factor. As NH_4^+, behavior between Na^+ and Mg^{2+}	Protein metabolism	Cytoplasm nucleus, vacuole
P	PO_4^{3-}, HPO_4^{2-}, $H_2PO_4^-$	Phytin, phosphatidylcholine (lecithin), nucleoprotein	PO_4^{3-} often the limiting factor	Carbohydrate metabolism. Formation of bones	Mainly in the nucleus, as inorganic phosphate
S	SO_4^{2-}, HS^-, S^{2-}, H_2S, etc.	Cystine, cysteine, glutathione, chondroitin, taurine, etc., mustard oils	SO_4^{2-} only for higher plants. Fungi and bacteria also H_2S	Regulates oxidation and reduction?	?
Br	Br^-	Organic & inorganic?	"élément concentrateur".	Hormonal (e.g., physiology of sleep)	?
I	I^-	Diiodtyrosine and tyroxine. Also inorganic	"élément concentrateur" Often the minimum factor	Hormonal (animal development)	?
Cl	Cl^-	Inorganic?	?	? dispensable for most plants	Cytosol, blood serum
Na	Na^+	Ionized?	Increases permeability of protoplasm. Non-toxic for most organisms. Effect countered mainly by Ca^{2+}	Indispensable for a few select groups of living creatures	Blood serum, cytosol
K	K^+	Ionized	As Na^+, but more toxic. Detoxified by Na^+, but mainly by Ca^{2+}	The most important biogenic metal, catalyst?	In plasm, not in the nucleus
Mg	Mg^{2+}	Ionized? Or in protein and chlorophyll	Acts as an intermediary between Na^+ and Ca^{2+}. Can be detoxified by Ca^{2+}	Required for green plants	Plastids, plasm?
Ca	Ca^{2+}	Inorganic or with organic acids in phosphatidylcholines, etc.	Decreases the permeability of the protoplasm. Detoxified by K, Na or Mg	According to many, highly influential on the nitrogen cycle. Formation of bones	As phosphate, sulfate, carbonate, oxalate. In cytoplasm and vacuole. Often in the membrane
Fe	Fe^{2+}, rarely Fe^{3+}	Ionized or complex bond (masked iron)	Minimum factor?	Component of the red blood coloring (hemoglobin), plays a role in oxidation and reduction	Nucleus and plasm?
Cu	As ion	Complex bound	Minimum factor? Lethal in very low concentrations (oligodynamic effect)	In hemocyanin (blood color in many lower animals). Stimulates the formation of blood in higher animals	?

Figure V.6 Oxidation of black mud from a freshwater lake (Searsville Lake, California). Photo by the author.

microscope slide (Figure V.6), only the sulfide under the edges of the slide will be oxidized by the entering oxygen. On these edges the reaction of $Fe^{2+} \rightarrow Fe^{3+}$ will continue for quite some time, and we indeed also find the iron bacteria in that location (Figure V.6, white area by the arrow). On the border with the non-oxidized area (black) and the grey contour we find the colorless aerobic sulfur bacteria, which oxidize the hydrogen sulfide and obtain the energy necessary for their carbon assimilation. According to Deines (40), part of the hydrogen sulfide is oxidized to sulfur dioxide in, or outside of, the cell. The so-called sulfur "droplets" that can be found in or outside the bacterium consist of hydrogen polysulfide (H_2S_x).[19] Purple and green sulfur bacteria develop in the area above the black mud where there is no oxygen, but light is available. Here it is the chemical environmental factor, hydrogen sulfide, which determines the composition of a community of various groups of organisms.

A similar characteristic chemical environmental factor can also be found in other cases, but almost nowhere is it as clearly expressed as on the black mud.

---

[19]In fact, the droplets are elemental sulfur.

EDITOR'S NOTES

There is much to savor in this chapter. Again, following the geobiological theme of the book, Baas Becking puts organisms solidly in the context of the chemical environment. Indeed, he discusses both how organisms respond to the chemical environment and how they control it; this is still an active research area in contemporary geobiology and microbial ecology. In the first part of the chapter Baas Becking focuses on gases and relays astute observations about the role of photosynthesis in influencing the pH and oxygen content of natural waters including tidal mudflat communities. These observations were spectacularly confirmed with the development and application in the late 1970s and 1980s of microsensors for oxygen and pH, revealing, in some cases, oxygen levels of up to 1 bar pressure and pH levels of > 9.5 over depth scales of a mere a millimeter or two. Baas Becking correctly notes that high pH reduces the content of aqueous CO_2 in the water, challenging photosynthetic organisms to obtain this vital substance. However, this is not universally true, as more recent research has shown that many types of aquatic photosynthetic organisms can actively transport bicarbonate into the cell, converting it to CO_2 for their use.

Baas Becking offers a fascinating discussion on the function and requirement of major and minor elements for life. He correctly identifies the major elements, and many of the minor elements, used by cells. He also properly identifies the functions of many of these during cell metabolism. Particularly with respect to the minor elements, he notes that they are not universally required, but that it depends on the organism, and speculates further that many of the minor elements have functions that are not yet known. Indeed, major improvements in trace metal analytical techniques, as well as protein structural analysis, have allowed a vastly more complete understanding of the role of trace constituents in the life process, but reinforcing in general what Baas Becking offers here.

Baas Becking argues that many of the constituents of life are limiting in nature, and that these limiting nutrients, when supplied in greater amounts, will increase the growth rate of a population until a certain point where the nutrient becomes saturated and has no further effect. This, in words, was formulated somewhat later in equation form by Jacques Monod, giving us the well-known Monod kinetics, which are still a standard way of analyzing the influence of limiting substrates on growth. All in all, the role of limiting elements on growth is still an active microbial ecological and geobiological topic. There has been much recent and exciting research devoted to exploring how trace elements and nutrient limitation have evolved through time, in pace with the evolving chemistry of the surface environment. Thus, patterns of trace element and nutrient limitation have likely influenced both the activity levels and the evolution of various types of aquatic organisms through time.

Finally, Baas Becking puts us in the chemical environment of black mud and discusses how the redox interfaces generated by the mud itself, and its interaction with oxygen, provide the chemical conditions for the development of a wide range of specific types of microbial metabolisms. Indeed, all of these different microbial types originate from the same piece of mud, yet dominate depending on the chemical circumstances of how the mud interfaces with the environment. "Everything is everywhere: but the environment selects"! With better analytical techniques for exploring chemical gradients in nature, the interrelationship of organisms with these gradients is an active geobiological topic today. Indeed, this is such an active research area that this topic might have dominated a whole chapter if Baas Becking were to rewrite the book today.

CHAPTER VI
Cycles

Everything flows; nothing is permanent. Yet our perception of the world is often a static one. A silent forest, a clear pond – deceptive snapshots that conjure for us the sense of inertia. However, natural processes constantly renew themselves in time and space. The symbol of the circle, of the snake with its tail in its mouth, of the cycle, represents this rhythmic recurrence.

Astronomical cycles – the tide, day and night, the phases of the moon, the changing of the seasons, the periodical return of comets – have long been proof of a predictable recurrence. The life cycle of a butterfly or of a higher plant, though different in nature from our own, have been known to humans, at least in general terms, for centuries. We also see the sequence of the forms of a certain matter, namely water. Therefore, even with little scientific knowledge, people have been able to recognize the water cycle as a third type of cycle.[1]

However, when we wish to describe the chain of events undergone by a certain substance, such as carbon, we encounter the countless organisms that carry this carbon in one form or another, absorbing or excreting it in order to capture or release energy. In such a cycle the energetics are of importance. Obviously, this type of cycle is much newer to us than the others, as the transfer of energy only became known in the mid-nineteenth century.

The second type of cycle mentioned above, that of elements or bonds, is important to environmental science when these elements or bonds are important for life. This is the case with most elements that occur in the "vadose zone"[2] (the so-called upper lithosphere, or biosphere). These elements are for a

---

[1]"From the Heavens it comes, to the Heavens it must return" (Goethe). [Original footnote from Baas Becking.]
[2]The vadose zone is normally considered the unsaturated zone between the land surface and the water table.

Baas Becking's: Geobiology, Or Introduction to Environmental Science, First Edition. Edited by Don E. Canfield.
© 2016 John Wiley & Sons, Ltd. Published 2016 by John Wiley & Sons, Ltd.

large part "cyclical," meaning that they can occur in many forms, converting back and forth between them. Vernadsky (120) characterizes a cycle as:

> *Each element has characteristic combinations that ensure its constant renewal. After the changes, more, or less complex elements return to their initial forms, beginning the cycle anew, which again ends by the element assuming its initial form.*[3]

Most cyclical elements are also involved with life. Examples of cyclical elements are sodium (which will be further discussed in Chapter VII) and iodine, of which the function as a cyclic element has been thoroughly studied, especially lately in its relation to goiters[4] (75).

Figure VI.1, taken from the Struma report, shows the iodine cycle, which, according to recent research by van de Eds (48), is also valid for fluorine.

There are phases in such cycles whereby the element is diffusely distributed (e.g., iodine in the ocean), alternating with phases in which it is present in concentrated form, such as with iodine in seaweed and sponges and in the thyroid glands of higher animals.

Seeing as how cycled elements are transported through water, the cycle of water is doubly important for the biologist. Lotka compares the water cycle on Earth with a Soxhlet extractor,[5] whereby the lithosphere represents the

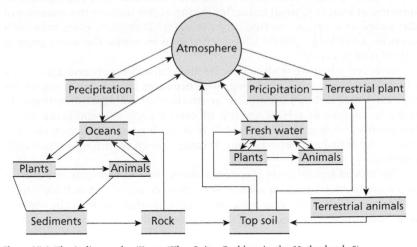

Figure VI.1 The iodine cycle. (From "The Goiter Problem in the Netherlands.")

---

[3]Original French: "*Chaque élément donne des combinaisons charactéristiques qui se rénouvellent constamment. Après des changements plus ou moins complexes l'élément revient à la combinaison première, et recommencera de nouveau un cycle qui, pour l'élément, se termine par son retour à l'état initial.*"

[4]An enlargement of the thyroid gland.

[5]A Soxhlet extractor a wonderful piece of organic chemical glassware by means of which solvent boiling in a flask is condensed into a chamber holding a sample for extraction. When the chamber is filled, the extracted solvent is siphoned into the solvent reservoir, where fresh solvent is then distilled and condensed again into the sample reservoir.

compound that needs to be extracted and water is the solvent, of which the vapor is repeatedly condensed [as rain] and reused for further extraction.

Among the biological cycles, we have already mentioned the "cycle of life" and its combination with chemical and/or energetic factors. As long ago as 1775, Joseph Priestley (102) identified the plant–animal cycle. J. Pringle (101) conducted a more thorough examination of this cycle in his *Eulogy on Priestley for the Royal Society*. J. V. Liebig (1840, 77), and later J. B. Dumas & J. Boussingault (1844, 47) further elaborated on these ideas, portrayed as the "cycle" in Figure VI.2. Presented here is the idea that chemical energy can be seen as potential energy until it is released and expressed as kinetic energy.

The carbonate molecule lies like a stone in a field, with an "energy level" of 0. Energy from solar radiation increases the energy level to that of sugar, at which 114,300 cal. per mole are absorbed.

Sugar then represents a certain amount of potential energy (674,000 cal. per mole) which can be transformed by the living cell into the power of motion. The metabolism of the cell consists of an "anabolic" part, whereby kinetic energy is transformed into potential energy, and a "catabolic" part, where the opposite is the case.[6]

Justus von Liebig has thoroughly reviewed the anabolic–catabolic relation between "green" and "colorless" life and has symbolized it in an aquarium. This microcosm, which we can call "Liebiger World," was very popular in

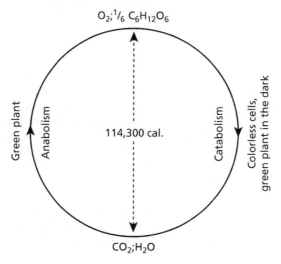

Figure VI.2 The basis of the cycle.

---

[6]We normally think of anabolism as the energy used to build biomass and catabolism as the energy used in metabolism. These views are consistent with those expressed here by Baas Becking.

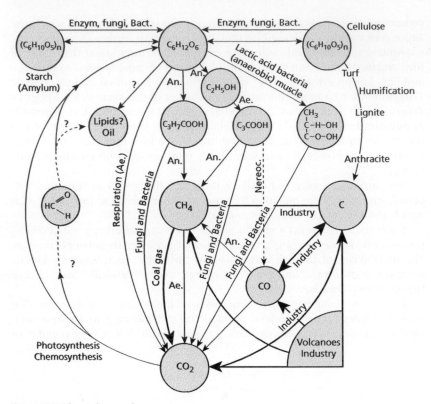

Figure VI.3 The carbon cycle.

Munich in the 1840s: an equilibrium between forces that break down and build up, between kinetic and static, between plant and animal.

We will briefly discuss the elements whose cycles are closest to living systems, namely carbon, nitrogen, and sulfur (Figures VI.3, VI.4, and VI.5).

The starting point here is the form of an element in its high oxidation state, for instance CO_2, NO_3^-, SO_4^{2-} [placed at the bottom of the figure]. Furthermore, each cycle has placed in its center an element in its lowest oxidation state, e.g., CH_4, NH_3, H_2S.

The elemental form is shown in the right-hand side of the cycle, while the more complicated forms are placed at the top. The location on the circle roughly compares to the amount of potential energy that is locked within the bonds.

The non-biological (geochemical and industrial) processes are depicted with thick lines. We now begin exploring the *carbon cycle* (Figure VI.3). Here a simple sugar ($C_6H_{12}O_6$) stands at the highest level; a sugar, which perhaps is formed from carbon dioxide (located at the bottom) via formaldehyde using chemosynthetic and photosynthetic processes. In order to raise a carbon dioxide molecule to the "sugar level," an input of energy is needed. The transformation of a simple sugar into complex carbohydrates such as starch and cellulose takes place in green plants. These changes take place without much exchange of energy, depicted in the diagram as the connecting horizontal line.

Another cell-wall component, lignin (wood polymer), and perhaps also cellulose, form products increasingly richer in carbon during the humification process [as these compounds decompose in soils and natural waters]. In the end, pure carbon is formed.

This is probably also possible in a sterile environment, without the help of microbes. This carbon is also expelled to the atmosphere in large quantities in the form of soot by industry. Although the oxidation of carbon to carbon dioxide provides abundant energy, we have not yet found any organism that can profit from this energy; the "carbon bacteria" have not yet been discovered!

To elaborate on the breakdown of sugars by living cells is beyond the scope of this book. Fermentation and aerobic respiration are only briefly discussed here, while several major types of metabolisms (butyric acid, alcoholic, and lactic acid fermentation) have a place in the diagram. Here "An." means anaerobic, "Ae." an aerobic process.

The formation of acetic acid and the formation of biogas (methane) are considered next. The methane bacteria [methanotrophs], which obtain the energy they need for their carbon dioxide assimilation from oxidation of the CH_4, were mentioned in Chapter II.

Carbon monoxide, which plays a role in industry, rarely occurs in living organisms, yet the "swim bladders" of a large type of kelp (*Nereocystis luetkeana*) contain a large quantity of CO that is possibly formed from volatile organic compounds (formic acid?). It is curious, though, that Beyerinck was able to isolate an organism that can obtain its energy from the oxidation of CO to CO_2 and that there is sufficient CO present in laboratory air to obtain good-quality cultures of this "*Bacillus oligocarbophilus*" on a substrate consisting of tap water to which a small amount of phosphoric acid has been added. Recently an organism was found in sewage sludge, which is able to reduce CO to CH_4, and a way has been discovered to use this organism to turn the previously mentioned (harmful) compound [CO] into useful methane.

Strictly speaking, the carbon cycle contains only a few chemosynthetic autotrophs, namely the forms that oxidize CO. Based on geochemical grounds (i.e., its presence in volcanic gases), the author would prefer to consider CH_4 as also inorganic,[7] by which the methane bacteria appear as the second

---

[7]In line with Baas Becking, most would view the oxidation of methane with oxygen as a chemosynthetic process. Indeed, these organisms also utilize CO_2 fixation to make cell carbon. Also, in a process apparently unknown to Baas Becking, an important class of methanogens reduce CO_2 with H_2 ($CO_2 + 2H_2 \rightarrow CH_4$), and these are also chemosynthetic organisms.

chemosynthetic component in the carbon cycle. Methane fermentation, through which biogas is formed from organic acids, is a very natural occurrence in our canals and ditches.

We now look at the *nitrogen cycle* (Figure VI.4). The building blocks of protein, the amino acids, are placed towards the top.

From this protein or these acids, the plant is able to build complicated alkaloids (e.g., atropine, noscapine, codeine, etc.) as well as glucosides containing nitrogen (e.g., amygdaline). Many amino acids are decarboxylated through the work of microbes, whereby CO_2 is split off and amines are formed (RNH_2). Many transformations of amino acids lead to ammonia, which can also be formed from urea by uric bacteria. Ammonia can also be formed from fatty materials containing nitrogen (such as lecithin). Either way, many decomposition processes (catabolism) lead to ammonia, which then can be oxidized to nitrite, and later to nitrate, each step producing

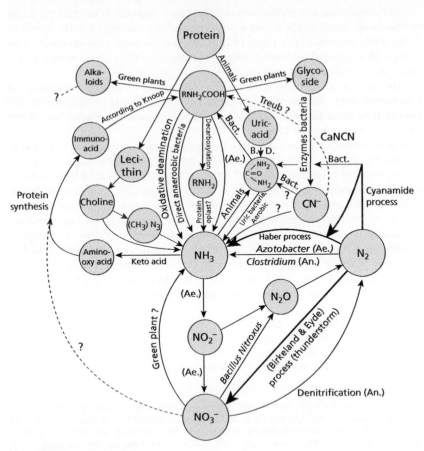

Figure VI.4 The nitrogen cycle.

energy. This energy can be utilized by chemosynthetic bacteria: namely the nitroso[8] and the nitro bacteria,[9] both discovered by S. Winogradsky. The nitrate formed is the nitrogen fuel *par excellence* for many green plants, especially in acidic environments. Here it is likely that nitrate is first transformed into ammonia [in the plant]. Reduction of nitrate takes place under anaerobic conditions. The energy level is increased; therefore the accompanying organisms are heterotrophic.[10] Products of this reduction can include laughing gas (N_2O), nitrite, and (primarily) free nitrogen. This nitrogen (N_2) can be fixed under both aerobic and anaerobic conditions as well as by heterotrophs.

The aerobic process [of nitrogen fixation] is performed by bacteria of the genus *Azotobacter*, mainly found in locally elevated concentrations of sugar.[11] The anaerobic fixation of nitrogen is performed by butyric acid bacteria: *Clostridium pasteurianum*. The technical Haber process does the same, while the Birkeland & Eyde process, based on the oxidation of atmospheric nitrogen [to nitric acid], has no biological analog. There are several theories regarding the synthesis of amino acids from simple molecules such as NH_3. A well-known idea comes from Melchior Treub, who assumes that HCN occurs first, which, through a reaction with ammonia and later saponification, produces the amino acids. However, it seems that the cyan in the leaves is most likely produced by the splitting of glycosides. Furthermore, there are many other objections to Treub's theory, and the cyanide synthesis [idea] is only of historical significance. According to Knoop, the amino acids are formed from ammonia when these bind with keto acids (e.g., pyruvic acid), from which an aminooxy acid would be produced, which after several simple conversions can be transformed into an amino acid. The nitrogen cycle thus offers us two chemosynthetic autotrophs, namely the nitrite and nitrate bacteria.[12]

Many more autotrophs can be found in the *sulfur cycle* (Figure VI.5).

One of the most complicated (and also most common) forms in which sulfur appears in living systems is the amino acid cysteine and its oxidation product cystine. We also find taurine as a protein component. These complicated bindings can produce a number of simple sulfur compounds

---

[8]Conducted by both ammonium-oxidizing Bacteria (AOB) and ammonium-oxidizing Archaea (AOA).

[9]Conducted by Bacteria, mainly of the groups *Nitrospira* and *Nitrobacter*.

[10]Within the last 15 years, the autotrophic process of anaerobic ammonium oxidation (anammox) has also been discovered to produce N_2 gas.

[11]Subsequent work has shown that nitrogen fixation in not an aerobic process, and indeed, the nitrogenase enzyme, which promotes nitrogen fixation, is deactivated by oxygen. When nitrogen is fixed in the presence of oxygen, the nitrogen fixating organism adapts one of many strategies to shield the nitrogenase enzyme from oxygen.

[12]The anammox bacteria offer another, in this elegant reaction: $NO_2^- + NH_4^+ \rightarrow 2N_2 + 2H_2O$. There is also a recently discovered autotrophic interaction between nitrogen and methane: $8NO_2^- + 3CH_4 + 8H^+ \rightarrow 4N_2 + 3CO_2 + 10H_2O$, as well as autotrophic microbial interactions between nitrogen and sulphide: $NO_3^- + H_2S \rightarrow S^\circ, SO_4^{2-} + N_2, NH_4^+$.

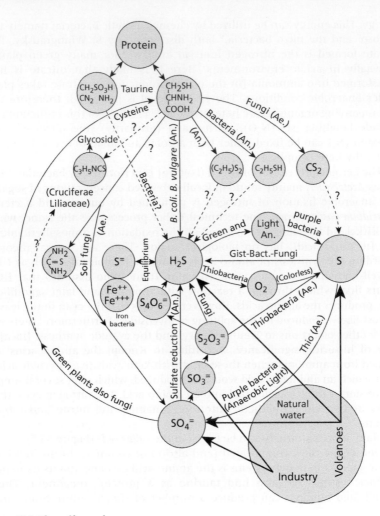

Figure VI.5 The sulfur cycle.

through bacterial activities (rotting), such as mercaptan and thiols as well as H_2S. It is also known that a fungus (*Schizophyllum*) is able to produce carbon disulfide. Until not so long ago, our textbooks taught that the rotting of protein was the main source of hydrogen sulfide; it is now known from research conducted by Beyerinck and van Delden that most H_2S in our canals and ditches is produced by the anaerobic heterotrophic reduction of sulfates. The H_2S created is a great starting point for further bacterial development. In the first place, it does not remain in its current form, but for the most part is bound to black hydrated ferrous sulfide, a commonly distributed matter found in black mud, described by A. F. Rogers as a specific mineral: hydrotroilite.[13] When this mud comes into contact with the oxygen in the air, the ferrous iron is transformed into ferric iron and bog iron; $Fe(OH)_3$ is formed. Sufficient energy is released during this process that specific chemosynthetic bacteria (the so-called iron bacteria) can develop (see Chapter VIII). H_2S can also serve as an energy source directly, as it oxidizes to sulfur when exposed to the air. The colorless, aerobic sulfur bacteria obtain their energy from this reaction as well as from the oxidation from sulfur to sulfate. Among these organisms are some which can also use thiosulfate as an energy source; these are described in further detail elsewhere in this book.

It is different when the H_2S does not have contact with air. Then the reaction $H_2S \rightarrow S \rightarrow SO_4^{2-}$ is only possible when light is absorbed, as it is with the colored sulfur bacteria which complete this photosynthetic process. The purple bacteria can transform the H_2S to sulfate, while the green bacteria bring it no further than sulfur.[14] The knowledge we have of this process is thanks to van Niel. We know that many organisms are able to produce the necessary [organic] sulfur bonds using sulfates, and even that other sulfur forms are generally toxic for them, though we still do not understand this process. The thiocyanates that occur in some mustard oils are more likely to be waste products than intermediate products for possible further metabolism.

The contributors to the cycle not involving life are primarily sulfates (in natural bodies of water), as well as H_2S and S from volcanoes and SO_2 which occurs in industrial gases and can inflict considerable damage (smoke damage) to crops.

Vernadsky says the following on inorganic cycles: "These cycles are perfectly reversible."[15] This is certainly not the case for biological cycles.[16] Nonetheless, there are stages here which disable certain bonds, sometimes for long periods of time. These substances could be called "snails of the biological industry."[17] The following table provides an overview of these products.

---

[13]This is true, but we now know that pyrite (FeS_2) is the major repository of sulfide in most anoxic sediments.

[14]It is now known that sulfate may also be an oxidation product, particularly when sulfide is limiting.

[15]Original French: *"Ces cycles sont parfaitment réversibles."*

[16]One could argue whether this statement of Baas Becking is true or not. Indeed, almost all biological processes processing matter (organic and inorganic compounds) seem to be reversible, not by individual organisms, but by the ensemble of organisms making up the global ecosystem.

[17]This is a fascinating discussion, in which Baas Backing is differentiating between the fast cycles of biological turnover and the slower cycles of geological turnover. The latter, in a sense, are specific examples of the types of cycles alluded to by Vernadsky, and by Hutton 150 years before him.

The "snails" in the cycle of living organisms	
I. Organic leftovers	Petroleum, asphalt
	Turf, lignite
	Coal, anthracite
	Guano, amber
II. Inorganic remains and pseudomorphs	Phosphate, e.g., vivianite
	Diatomite
	Limestone (from coral)
	Coraline algae (nullipore)
	Shells
	Pteropoda and Globigerinida calcium
III. "Matrix" operations of organisms	Bog iron (iron oxyhydroxide)
	Kieselgur (hot springs)
	"Photosynthetic" calcium
IV. Bacterial changes in the environment	Hydrotroilite → pyrite (?)
	Sulfur
	Sodium nitrate?
V. Stimulates sedimentation	Sediment deposits by *Cardium*
	$CaSO_4$ by *Artemia*

Direct organic remains can accumulate in dry climates (guano) or when cut off from oxygen (petroleum).

Some organic materials, such as chitin, are so resistant that they can still be identified in Paleozoic organisms (trilobites). The outer layer of a grain of pollen, consisting of the resistant material exinite, makes it possible for these grains to remain fully identifiable.

Inorganic remains – the completely mineralized remains of organisms – consist, in part, of phosphate minerals, which come mainly from bones (an example is the light blue iron phosphate mineral vivianite from our raised bogs). Many organisms accumulate silicates or carbonates in their cell walls or excrete them as (sea) shells. Diatomaceous earths, limestone from corals, which are found frequently in our country, also belong to this group. Mineral compounds can also precipitate on a living structure. This is the case with bog iron, which is often deposited around the mucous vacuole of iron bacteria, the kieselgur[18] which precipitates onto cyanobacteria in hot springs (Yellowstone Park); as well as the calcareous deposits on some freshwater aquatic plants. An inorganic product of the biological cycle (FeS, S, $NaNO_3$) can also remain in an external state, thus breaking the cycle.[19]

These circumstances can be extreme drought (sodium nitrate) or being closed off from air (e.g., sulfur in the clay of the Zuiderzee, or pyrite in deep layers of mud).

---

[18]Kieselgur is a silica oxide precipitate.
[19]The fixation of biological products into mineral form doesn't really break the cycle, but it introduces these products into the longer-term geological cycle.

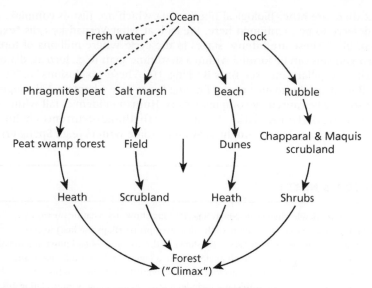

Figure VI.6 Succession of associations (after data from Walter).

Finally, I would like to point out the role organisms fulfill in the precipitation of clay or limestone. Richter has proven this for the cockle (*Cardium*), while the shrimp *Artemia* precipitates calcium from seawater concentrates (see Chapter X).

In the following chapters we will see that the cycles possess specific characteristics that depend upon the type of environment in which they play out. Indeed, a complete ecosystem is characterized by its cycle. In ecology, the branch of science that concerns itself with the environment and with the interrelations between organisms, these so-called "successions" may be viewed as cycles. Cowles, who first published on this topic in 1911, speaks of "vegetative cycles." Weaver & Clements (125) define such successions as follows: "As vegetation develops, the same area becomes successively occupied by different plant communities. This process is termed plant succession." The succession ends in a so-called "climax," which in our climate is almost always a forest. Figure VI.6, based on data from Walter (123), shows several of these successions. The climax itself, though of the highest order, is also not an end. Braun Blanquet (31) remarks on this: "The climax is also but a point of rest, a breather in the eternal process of vegetative succession."[20]

Having learned of succession, we have come to know one of the highest expressions of cycles.

---

[20]Original German: "*Auch der Klimax ist aber, das sei ausdrücklich betont, nu rein Ruhepunkt, ein Atempause im ewigen Wandel der Vegetationsgestaltung.*"

Yet there are other biological phenomena which are just as complex, and they deserve to be mentioned here. We look at what Vernadsky calls "explosions of life." These are intense stages in the cycle where millions of tons of living organisms can be formed within a short time span (e.g., locusts, diatoms, jellyfish, caterpillars, etc.; see Baas Becking, 10). These "explosions" of life can occur due to an abundance of the limiting factor (see Chapters 3 through 5); they also may be caused by complex factors. Human epidemics fall within this same framework (Lotka, Clarke, 38), as do the rhythmic occurrences in human history which have been brought forward by some writers (e.g., Spengler).

EDITOR'S NOTES

This is a remarkable chapter in which Baas Becking shows us what he means by geobiology. He gives us here some of the first attempts (perhaps the first) to show how bioactive elements form cycles. He shows us the coupling of all known microbial processes in the cycling of individual elements in a logical and clear manner, using exactly the same types of diagrams that are used today. In testament to his geobiological thinking, Baas Becking includes on the same diagrams interrelationships between the biological and abiological cycles, thus coupling the short- and long-term cycling of elements on the Earth.

He also understood, at least in general terms, how the biological cycles and geological cycles transferred elements between them. He states:

Vernadsky says the following on inorganic cycles: "These cycles are perfectly reversible." This is certainly not the case for biological cycles. Nonetheless, there are stages here which disable certain bonds, sometimes for long periods of time. These substances could be called "snails of the biological industry."

Geobiologists are still very much occupied with the quantitative relationships between the fast (biological) and slow (geological) cycles. Much of this understanding has emerged since the plate tectonic revolution showed us both how and how fast the Earth moves, and how this motion defines the cycling of rocks and the elements they contain. This understanding only emerged decades after Baas Becking wrote these pages. Since his time there also has been an accumulating understanding of the rates of biological processes on a global scale (e.g., primary production, sulfate reduction, nitrogen fixation, etc.), and this understanding has enabled a quantitative coupling between the fast and slow element cycles of the Earth.

In presenting his cycles, Baas Becking takes us through a wonderful tour of known biological metabolism. Indeed, they knew a great deal in 1934! However, we know so much more today, and much of this understanding has emerged only in the last 20 years. Many "new" microbial metabolisms have been uncovered, yielding relationships between the element cycles that were unknown in Baas Becking's time. To name a few, we know now that the sulfur and carbon cycles are coupled through the anaerobic oxidation of methane with sulfate, and the anaerobic oxidation of methane with nitrate couples the nitrogen and carbon cycles. An autotrophic metabolism in the nitrogen cycle, the anaerobic oxidation of ammonium with nitrite (anammox) has recently been uncovered, and it is of global significance. There is also a whole biological cycle involving iron that was unknown in Baas Becking's time.

Finally, in this and other chapters, Baas Becking pays homage to the great Russian "geobiologist" Vladimir Vernadsky. It is clear that Vernadsky was of great influence to Baas Becking, and if this book, *Geobiologie*, was translated into English much earlier, Vernadsky's name would have likely stayed alive among English-speaking Western scientists. Happily, Vernadsky is reemerging as an important trail blazer in our appreciation of life as a geological force.

CHAPTER VII
Oligotrophic Water

Alongside the infusions of science are those of nature.

M. W. Beyerinck

Natural bodies of water are characterized by the quantity and the nature of particles dissolved within them. If there are few dissolved particles (e.g., less than 100 mg/L), the water is considered to be *oligotrophic*. In contrast to this is *eutrophic* water, which contains more particles. The comprehensive characterization of the ecosystems in these waters is subject to the science of *limnology* (117). Seawater, a solution with a remarkably regular composition, together with its ecosystems, is described by *oceanography* (58). There are also several environments which do not fit into the above-mentioned categories, and some of these can be described as *dystrophic*, meaning less suitable for life. As of yet we do not have the data to be able to judge whether a natural environment should be considered as more or less "suitable," which perhaps makes the choice of this term [dystrophic] somewhat awkward. It is certain, however, that the biological characteristics of a body of water – its ecosystem (biocenosis[1]) – are often determined by its chemical composition.

For an oligotrophic ecosystem, water with very little dissolved matter is required. One source of this type of water is precipitation. The matter transported through the atmosphere was discussed in Chapter II. Precipitation contains, whether dissolved or in suspension, sometimes quite large quantities of the most varying substances. The environments (biotopes, or places where biocenosis can develop) which develop from this rainwater are characterized by the nature of this precipitation.

Numerous analyses of rainwater and snow have been conducted. Important summaries can be found in the works of Bunte and Clarke (34,36), for example. From these it can be noted that the consistency of this water is quite variable, as can be the amount of dissolved matter (15–70 mg/L). Here are some examples:

---

[1]An alternative word for ecosystem, not widely used in English.

Baas Becking's: Geobiology, Or Introduction to Environmental Science, First Edition. Edited by Don E. Canfield.
© 2016 John Wiley & Sons, Ltd. Published 2016 by John Wiley & Sons, Ltd.

	Fécamp (1850)	Centr. Lab. Utrecht Wijster (Drenthe) March 1933
Na^+	11.9	4.9
Ca^{2+}	5.8	3.3
SO_4^{2-}	11.1	7.5
Cl^-	10.3	4.9
NH_4^+	1.0	—
NO_3^-	0.9	—
Mg^{2+}	—	0.7
Organic	9.0	—
Total	50.0	21.3 mg/L

Sometimes the amount of organic matter can be much greater (up to 80% of the total dissolved matter). The ammonia content is often high in cities – higher in wintertime (3.76 mg/L) than in summertime (2.58 mg/L, Cirencester, 26-year average) – while urban fog can also contain large amounts. Usually the ammonium content is reduced after several rain showers. In a sample of rainwater collected in Leiden in the spring of 1933 we found 5 mg of NH_3 per liter. Based on the annual average of 3 mg/L, the amount of ammonia which enters the soil through precipitation is, at an annual rainfall of 75 cm, at least 22 kg per hectare. In the tropics rain contains nitrogen in the form of nitrate. Pierre calculated that in Caen the following minerals were deposited in the soil (kilograms per hectare per year):

NaCl	37.5
KCl	8.2
$MgCl_2$	2.5
$CaCl_2$	2.8
Na_2SO_4	8.4
$CaSO_4$	6.2
K_2SO_4	8.0
$MgSO_4$	5.9
Total	79.5 kg/ha/year

All these substances must be attributed to mineral cycles, and in locations where evaporation is higher than precipitation, salts can accumulate (see Chapter X, Sambhar Lake). The mineral cycle can also be identified in places where precipitation is higher. There appears to be a relation between the concentration of chloride ions in surface water and the distance to the ocean, wind direction, and topography, in such a way that it is possible to construct "isochlorides" (lines which connect the points at which similar levels of chloride ions can be found in the groundwater). In many cases, these lines follow topography, as in the eastern part of the United States and on the San Francisco peninsula. Figure VII.1, derived

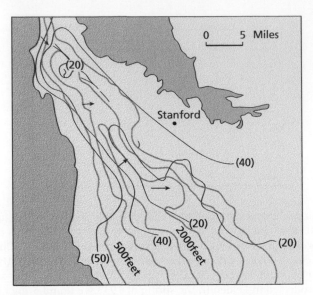

Figure VII.1 Map of the San Francisco peninsula, California, with isochlorides marked. (From J. P. Mitchell)

from research by Mitchell (86), shows the isochlorides (in mg Cl/L) for surface water south of San Francisco. From the map one can see that the [prevailing] western wind, the distance to the ocean, and the topography are highly influential. The arrows highlight the points where ocean fog travels inland (fog channels). As far as the author is aware, there are no such systematic data available regarding the amount and nature of dissolved matter in rainwater. The Struma Commission in the Netherlands has done pioneering work on iodine (75).

From the previous section it is clear that a biotope fed by precipitation receives a large number of substances from this water. Such a biotope is thus shaped by rain, and is known as an *ombrogenous* biotope. In contrast, another type of oligotrophic biotope is the *soligenous* type, in which water leaches mineral-poor soil. The water from snow melt and upstream in some rivers is often also extremely oligotrophic. In mineral-poor, fast-flowing river water, the supply of nutrients seems to be sufficient due to the speed of the current.

This is why in well-aerated oligotrophic waters the diversity of organisms depends on the stream velocity. The river water regularly acquires minerals from the rock bottom, so that the river downstream can be eutrophic (Truckee River, California, for example).

We will limit ourselves here to another type of oligotrophic water: raised-bog water. The reader needs to realize that an ecosystem such as this cannot be completely described in such a short discourse. For further detailed descriptions we recommend the works of Harnisch (57) and W. Beyerinck (25). In contrast to oligotrophic river water, bog water is stagnant and still. There is no dynamic contact with the atmosphere and, because diffusion of gases in water

Figure VII.2 Leaf of *Sphagnum recurvum* (greatly enlarged) with hyaline and green cells. Bacteria can be found in the hyaline cells. (From a sketch by Ms. A. v. Oven)

is a very slow process (Chapter V), the gaseous equilibrium is in this case biologically determined. Photosynthesis by the submerged (underwater) plants is not able to counter the oxygen-absorbing processes, and thus raised-bog water is characterized by low oxygen levels. We will return to this point later. Bog water is *ombrogenous* (or *soligenous*) – stagnating. The sealing off of the sub-soil in our region is caused by non-soluble matter originating from the bog itself. As little is known about the nature of these humus-like products, we will not delve any further into this matter.

The Dutch raised bog is mainly the product of one characteristic plant genus: *Sphagnum* or peat moss. These remarkable plants occur in a large variety of species, which all have the same structure. For our purpose we are inter-ested in the cells of the leaves that sprout on the side shoots and along the stem. The living leaves contain two kinds of cells: elongated cells with living matter and a large number of chlorophyll granules; and lifeless, colorless cells, supported internally by rib-like structures, connecting to the external environ-ment through pores (so-called hyaline cells). We pointed out in Chapter V that *Sphagnum* absorbs moisture through its leaves. The living cells are able to absorb water through a large surface area from the hyaline cells. These lifeless cells harbor a whole ecosystem, including especially the "sphagnicole" *Rhizopoda*, but also many types of bacteria. Figure VII.2, drafted after a sketch by Ms. A. v. Oven, shows the two types of cells in the leaf of submerged peat moss, *Sphagnum recurvum*, with bacteria in the colorless cells. With the excep-tion of the submerged peat mosses, most are already dead several centimeters below the water surface. The bog layer that is formed amounts to a few mil-limeters per year. Peat mosses can grow in such a way that the entire bog is

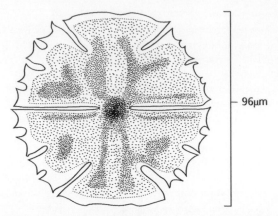

Figure VII.3 *Micrasteria truncata* (Corda) Bréb., a Desmidiaceae from raised bog water. (After a drawing by K. Vaas)

elevated several meters above the original level, creating a "head" of living material on the surface which towers over the surrounding area, and only its strong water-adhesive power prevents it from drying out. In this case there is no doubt that the whole living mass is constructed from the atmosphere (in the broadest sense of the word).

Regarding the various types of peat moss, the systematic, ecological, and biological peculiarities were recently extensively described (25). The other life forms which inhabit the "*Sphagnetum*" will be mentioned here only briefly. The higher plants we encounter live symbiotically with fungi in order to effectively make use of the nutrient-poor environment (Ericaceae, such as *Calluna, Erica, Vaccinium, Andromeda*). Other higher plants seen here are crowberry (*Empetrum*) and inundated club moss (*Lycopodium inundatum*). Sometimes the hunger for nitrogen is satisfied by insectivoria, such as *Drosera* and *Utricularia*. Noteworthy, besides the peat moss, are pincushion moss *Leucobryum* and the submerged liverwort *Lophozia inflata*.

Regarding the algae, it can be stated that the diatoms have been strongly reduced [in numbers], while many common green algae do not occur at all in this environment (*Cladophora, Vaucheria,* etc.). A particular group of conjugate algae (a subgroup of green algae) – namely the *Desmidiaceae*, which stand out due to their elegant shape (see Figure VII.3) – attain their largest variety in form in just these waters, and dozens of species can be found here.

Other green algae are the characteristic *Oedogonium itzigsohni,*[2] while, oddly, a red alga (*Batrachospermales*) also occurs abundantly. Cyanobacteria and flagellates are found profusely. A number of these forms also populate snow melt and strong brines.

---

[2]This is apparently an outdated species name within the genus *Oedogonium*.

Fauna, with the exception of the previously mentioned *Rhizopoda*, is generally poor. Shelled *mollusks* are absent (due to calcium deficiency and high acidity), although a few remarkable crustacea can be found here (including *Eurycercus glacialis*). Of the fish, there remains only the stickleback, which we come across in many ecosystems.

The presence or absence of certain life forms cannot be explained solely by the lack of dissolved nutrients, or "nitrogen starvation." The high acidity of bog water is one of the largest factors determining which life forms can occur. In Chapter V, the high acidity of this environment (pH 3–5) was highlighted, as was the fact that many microbiological processes are influenced by the pH level. The observed traits of bog water can be attributed at least partially to its acidity. Grains of pollen and spores remain intact for thousands of years in this environment; the so-called bog people[3] also highlight the suppression or absence of certain microbial activities.

The cause of the high acidity levels is still not understood (16). It is known that bog water contains many different kinds of organic acids, and that it is rich in free carbon dioxide. Yet these amounts are too small to account for the high acidity. In the context of the classic work by Baumann & Gully (19) and also due to experiments done by K. Vaas in the Botanic Laboratory in Leiden, the author concludes that one of the largest contributing factors causing the acid formation is:

As with many plant cell membranes, the cell walls of peat moss have the capacity to absorb salt from metal ions and to exchange these for hydrogen ions. Peat moss in a highly diluted solution of ammonium chloride will soon acidify this solution, while the ammonium level of the solution is reduced and the chloride level remains the same during the experiment. Rainwater and diluted solutions of table salt and of copper chloride show analogous behavior. If one takes peat moss which has been rinsed and places this in distilled water, the acid content will not increase and can even decrease due to active carbon dioxide assimilation. This makes it probable that the *Sphagnum* itself expels little or no acid, but that it is capable of an active ion exchange. As has been said earlier, dead or lifeless cell walls can absorb the most varying cations (even copper).

The acid which is formed through the expulsion of H-ions depends on the nature of the environment. Experiments conducted by Ms. A. v. Oven in Leiden have provided further data regarding this environment. From these experiments it appears that the following bacterial processes are possible in the high acid levels of bog water:

Reduction of sulfates to sulfides; naturally without forming ferrous sulfide, seeing as how the acidity is too high for this and the amount of iron present too low.
Reduction of nitrate to free nitrogen.
Oxidation of sulfides (at the surface of the bog).

---

[3]These are individuals mummified for periods ranging from decades to millennia in bog water. They are especially common in northern Europe.

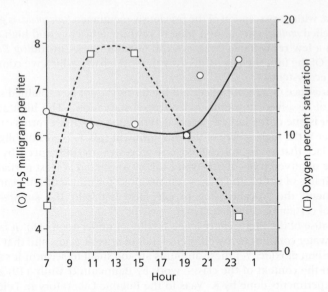

Figure VII.4 Levels of oxygen and hydrogen sulfide in bog water at different times during the day. (Measurements by Ms. S. Haspers and Ms. T. Hof. Wijster, Province of Drenthe, Aug. 1932)

Anaerobic formation of butyric acid from carbohydrates by typical butyric acid bacteria (*Plectridia*).

Purple and green bacteria, however, are absent. The acid levels are too high.

The presence of butyric acid bacteria may indicate possible anaerobic nitrogen fixation.

The presence of biogas in the bog water makes it plausible that methane fermentation occurs here as well.

The relatively large amount of sulfates brought in by rainwater, and the sulfate reduction which follows, causes the level of hydrogen sulfide in a raised-bog lake to often become quite high. The Leiden Biology Club was able to confirm this through a series of observations on an excursion to a raised-bog lake near the town of Wijster in August of 1932. With these field observations, both oxygen and hydrogen sulfide levels were determined (Figure VII.4). Oxygen concentration is expressed in percentage saturation, meaning the fraction of the whole that can be contained in water [at air saturation]. Oxygen increased in the morning hours and was at a maximum in the afternoon. Parallel to this, the hydrogen sulfide level declined. The increase in oxygen is a direct consequence of carbon dioxide assimilation and photosynthesis, while the decline in hydrogen sulfide is an indirect consequence of this. Although it is likely that there are other oxygen-utilizing processes taking place in the bog water, that of sulfate reduction is the most important.[4] Around midnight the oxygen level is

---

[4]Here, Baas Becking is referring to the oxidation of the sulfide formed from sulfate reduction.

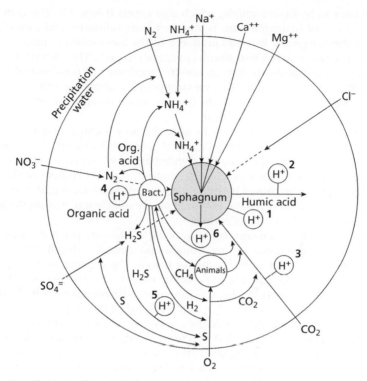

Figure VII.5 The cycle of matter in raised bog. Explanation in text.

extremely low, the H_2S level very high (7.5 mg/L). The high H_2S tension (4×10^{-4} normal[5]) in bog water, however, is not able to suppress the development of algae in the water. Higher concentrations are needed to attain that, as proven by Niel (10^{-3} normal).

Based on the above-mentioned data we can now conditionally depict the cycle of matter in a raised bog lake (Figure VII.5).

Here the circumference represents the atmosphere, with mineral components of the atmosphere[6] represented as points on the circumference.

Within the circle is the bog lake, and in the center the peat moss which absorbs the cations Na^+, NH_4^+, Ca^{2+}, and Mg^{2+} and which therefore expels H-ions (number 1 in the figure). Atmospheric nitrogen can possibly be bound by butyric acid bacteria when NH_4^+ is formed. Nitrates are reduced to nitrogen

---

[5]At the low pH of the bog water, all of the sulfide is the form of H_2S. Here, Baas Becking has given the concentration of H_2S in normal units which are usually used to express acid concentrations. With two exchangeable hydrogen atoms, the normal concentration of H_2S is twice its molar concentration.

[6]Here, Baas Backing is presumably referring both to gases in the atmosphere and to the dissolved constituents of rainwater.

gas, sulfates to hydrogen sulfide, which also expels H-ions (5). The carbonic acid, a source of hydrogen ions, is absorbed from rainwater and assimilated through the *Sphagnum*. It is also produced in the metabolism of bacteria and animals, the last of which also absorb [consume] oxygen. The bacteria are able to excrete organic acids (4), but can also free methane and possibly hydrogen, which might be oxidized in this environment by other bacteria. So-called humic acids are also produced in the bog (2), and it remains possible that the peat moss itself also excretes acids (6).

Because of intensive sulfate reduction, the only anion in bog water that remains practically unmodified by biological processes is chloride. From analyses of Wijster bog water, generously provided by Dr. W. Beyerinck, it appeared that the chloride content was 12 mg/L. Indeed, the acid content of the bog water can initially be caused by [the formation of] *hydrochloric acid*. The small amount of chloride in rainwater is quantitatively sufficient to explain the acidity, as can be seen from the following:

We assume that rainwater contains 5 mg of chloride per liter, bound as NaCl. This diluted solution is now brought into contact with cell walls which are capable of exchanging the Na-ion for H-ions. The concentration of H-ions will, in strong acid, be equal to the chloride ions; 1.38×10^{-4} normal. The pH amounts to $-(\log 1.38 - 4)$ or 3.86, which concurs with observations in the field.

As we saw previously, the high acid level makes the bog a very selective environment, in which many microbiological processes are inhibited. As the formation of raised bog is the first step toward carbonization, which leads first to lignite and later to coal, the thought is raised as to whether this process is caused by bacteria. The American bacteriologist Waksman (122), a propagator of the microbial nature of the carbonization process, points out that a large number of acid-tolerant, anaerobic bacteria occur mainly at greater depths within the peat. The degradation of cellulose is largely slowed, but the transformation of woody matter (*lignin*-like products) occurs with great intensity. According to recent investigations, it is precisely these lignin-like substances which are present in cell walls, which, after oxidation (*formation of humic acid*) and the further splitting off of carbonic acid, forms the so-called *humus matter*. Recently K. Griffioen (56) was able to prove that the black color in the core wood of ebony is caused by humus matter, most likely formed from the woody matter in the cell walls. It seems that this process takes place in a sterile environment (i.e., one free of bacteria). This would be an argument for the theory of sterile carbonization which was put forward by e.g., Bergius.

The lack of oxygen and the high acidity define bog water as much as the low mineral content. Small amounts of acid in eutrophic water could never cause a high acidity level, because of the buffering capacity of the water; while the amount of cations available for exchange in mineral-poor water can cause high acidity quite rapidly.

Also regarding the saline antagonism (Chapter IX), mineral-poor water holds a special place. Many organisms which generally can only occur in high abundance in so-called balanced saline solutions can also be found when "damaging" ions are absent.

Bog water represents an extremely complicated and extremely specialized biotope which inevitably will be formed where precipitation exceeds evaporation and where the atmospheric water either can be directly isolated from the eutrophic layer or is percolated through extremely mineral-poor soil. It is mainly through the research conducted by Dr. B. Polak (99) that we know that raised bogs can also be formed in the tropics when analogous circumstances are present, although it cannot be called a "Sphagnum biotope," a *Sphagnetum*. What we consider in our country to be raised bogs are but the last remains of what was once an expansive landscape. Several years ago, Dr. Polak (98) was able to prove that the Dutch "low bog" is nothing but raised bog which through subsidence has been submerged. The accidental conditions in our country are currently thus that we would be inclined to label raised bog as a special biotope. It is good to realize at this moment in time that all environments are equal and that ecology is not served by praising a certain biotope which, through coincidental situations, covers a large part of our research area or, if you will, of the Earth.

EDITOR'S NOTES

In this chapter Baas Becking takes us into the heart of geobiology. He describes the relationship between organisms and special "oligotrophic" environments which are poor in nutrients and low in primary production. The chapter focuses on those environments fed solely or mostly by rain and saves most of his fire for special raised-bog environments as found commonly in the Netherlands. These are seemingly close to Baas Becking's heart, and a type of environment studied by both him and his close colleagues. In this regard, he gives us a charming description of chemical analyses (highly relevant!) carried out on day excursions by the Leiden Biology Club. As usual, Baas Becking's observations are keen. He explains how the special low pH of these environments limits the populations of both microorganisms and macroorganisms (plants and animals) in the peat. He explains how nutrients and metabolic products such as sulfide are cycled, and provides a compelling view (courtesy of the Leiden Biology Club!) of the dynamics of oxygen and sulfide cycling over a diel cycle. Of special geobiological interest, Baas Becking also takes us from peat to coal and explores the role of both microbes and abiological processes in this transformation.

It is clear that Baas Becking has a special interest in raised peat bogs, but, ever the romantic (or so it seems), he argues that all ecosystems on the Earth have their importance and fascination, and that we should be careful of becoming overly enamored with our favorite or most convenient locations.

In placing the organisms squarely in the context of their environment, Baas Becking is extremely modern in his approach. The full descriptions of elemental cycles, and how they are regulated in different environments, is a common goal of modern geobiological (and microbial ecological) research, as is the approach of following the chemical dynamics of environments as they develop over diel time scales or even longer periods of time. In this chapter, one clearly observes geobiology as a scientific field.

CHAPTER VIII
Eutrophic Fresh Water

Limnology is the study of freshwater ecology.[1] The limnological literature is vast. If one wishes to discuss limnological questions in just a few pages, one must either limit himself to only definitions or focus on a well-defined problem. The chemical environment is thus an apt subject for this chapter.

Eutrophic fresh water can be defined as a solution containing approximately 200–600 mg of dissolved mineral salts per liter. If the solution contains more than 600 mg/L of soluble minerals, it is called brackish water, which will be discussed in the next chapter. In fresh water, the most common cations in order of quantitative importance are:

$$Ca > Na > Mg > K$$

and for anions:

$$CO_3 \left(+ HCO_3 \right) <> Cl > SO_4$$

In seawater, the sequence is the following:

$$Na > Mg > Ca > K$$

and:

$$Cl > SO_4 > CO_3 \left(+ HCO_3 \right)$$

We will use a simple graphic method to display the composition of these waters. This method is derived from an unpublished work by Dr. A. Massink as referenced by Ms. J. Ruinen in her dissertation (107).

A natural body of water can be characterized by a simplified solution, containing the cations Na (including K), Ca, and Mg, and the anions Cl (with Br, I, and F), CO_3 (with HCO_3), and SO_4.

---

[1]Most would also include freshwater (mostly lakes) chemistry and physics in the definition of limnology.

Baas Becking's: Geobiology, Or Introduction to Environmental Science, First Edition. Edited by Don E. Canfield.
© 2016 John Wiley & Sons, Ltd. Published 2016 by John Wiley & Sons, Ltd.

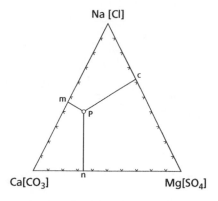

Figure VIII.1

Thus we have two groups, each having three components. With these, we are able to depict the relations between the cations and the relations between the anions by using points in a triangle diagram (Figure VIII.1). In the equilateral triangle Ca – Mg – Na a particular solution is portrayed by point P. The molecular relationship of Na, Ca, and Mg is such that the concentrations of Na : Ca : Mg = Pn : Pc : Pm. Due to a well-known property of the equilateral triangle, Pn + Pc + Pm = constant. Point Ca represents a solution in which the only cation present is calcium (100% Ca). Now the distance is Pc = 48, Pn = 38, and Pm = 14. The ratio of Na : Ca : Mg = 38 : 48 : 14.

Consider the same triangle with the corner points $Cl - CO_3 - SO_4$.

P now represents a solution in which the ionic ratio of Cl : CO_3 : SO_4 is 38 : 48 : 14. The composition of natural water can now be depicted by two points on a triangular diagram; one point (K) represents the ratio Na : Ca : Mg and the other point (A) the ratio Cl : CO_3 : SO_4. Figures X.1 and X.2 in Chapter X give certain characteristics for two types of river water, one type being CO_3 > Cl > SO_4 and the other Cl > CO_3 > SO_4.

Maucha (82) also recently designed a graphic method for the chemical description of waters, derived from the work done by I. Telkessy. The advantage of this method is that more than six components can be incorporated; the largest drawback is that it cannot be overviewed, which is the reason why this method will not [likely] be used.

The graphical method of describing water has many applications, of which some will be discussed in Chapter X. It has repeatedly been pointed out that various life processes can bring about chemical changes to the environment. One such change is depicted in Figure VIII.2, namely the two opposing processes of carbon dioxide assimilation and of respiration during sulfate reduction. When P represents the ratio of anions in a certain water body, the level of carbonic acid will decrease through photosynthesis and increase through respiration. The changes must be just so that the composition of water can always be represented by the points on the line $P - CO_3$, whereby it must be taken into account that CO_3 represents carbonic acid in all its various forms.

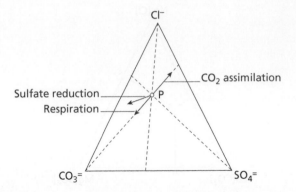

Figure VIII.2 Sulfaatreductie – sulfate reduction Ademhaling – respiration CO_{2-} assimilatie – CO_2 assimilation.

Sulfate reduced is simply the decrease in sulfate concentration.

The arrow which points to this process on the diagram should lie on the connecting line P – SO_4. However, in eutrophic water, sulfate reduction also produces sulfide,[2] which is a salt of weak acidity. There will thus be an alkaline reaction, meaning that the water will be able to hold more carbonic acid[3] A decrease in SO_4 is usually paired with an increase in CO_3, and the arrow which indicates sulfate reduction points toward $SO_4 \rightarrow P$ and $P \rightarrow CO_3$.

The solubility equilibrium of calcium carbonate can be exceeded through increased alkalinity, which can occur during carbon dioxide assimilation. This means that biogenic changes can also influence the ratio of cations.[4]

The low dissolved mineral concentration of river and lake water – the osmotic environment in the broadest sense – exerts great influence on living organisms. There is a great inclination for water *absorption* [into an organism]. The extent to which this occurs is probably due to the chemical nature of fresh water. Pantin (38) has been able to show that marine worms, when placed in fresh water, absorb water, but that this *absorption* of water changes into an *excretion* of water when calcium is added. Chapter V mentions the importance of ions in the cell's ability to absorb and excrete water – it is possible that the dominant role that calcium has in fresh water is of great biological importance.

---

[2]Baas Becking seems to imply that sulfide is not produced in non-eutrophic waters. More likely he means that sulfide is more likely to accumulate when sulfate reduction rates are high, producing more sulphide than there is Fe to capture it, and thus allowing sulfide to accumulate.

[3]The sulfate reduction reaction is: $SO_4^{2-} + 2CH_2O \rightarrow H_2S + 2HCO_3^-$. As displayed in the equation, the speciation of the sulfate and carbonate species are just an approximation of the speciation in natural waters of circumneutral pH. Indeed, in natural waters the pH of the solution will adjust, altering the speciation, such that the sum of charges represented by the sulfide and carbonate species added to the solution balance the charges lost from the reduction of sulfate.

[4]This is of course through the precipitation of calcium carbonate removing Ca^{2+} from solution.

Experiments conducted by Mr. J. Wakkie in the Botanic Laboratory in Leiden have given rise to the suspicion that even very small amounts of chlorides of potassium, calcium, and magnesium, when added to distilled water, exert a noticeable influence on the cells of green algae as well as of higher plants. The assumption, with respect to biological influence, that fresh water is equivalent to distilled water was disproved by Sydney Ringer in 1881 (see Chapter X). A careful analysis of fresh water is as desirable in environmental science as is the analysis of water in a culture of organisms. A slight increase in the NaCl level can have a negative effect on plant growth, while "hard"[5] water is an unsuitable environment for a large group of organisms. This "calcium shyness" is largely based on the damaging influence of low pH levels, and the increase in salinity must go hand in hand with osmosis, and thus the primary chemical influence of normal mineral components remains (Pearsall 94).

The so-called limiting factors such as phosphate and nitrate concentration have been studied by Atkins (5). In the shallow freshwater pools he studied, it seems that phosphate acts as the limiting factor for the growth of plankton, even more so than nitrate or ammonium. When phosphate disappears in the springtime, the further development of the plankton stops as well. In general, fresh water contains more phosphate (up to 400 mg/m^3) than seawater (up to 30 mg/m^3). The same is true for silicates. In fresh water this can amount to up to 3000 mg/m^3, while in seawater values of more than 250 mg/m^3 rarely occur.

The American limnologists E. A. Birge and C. Juday, who have been researching the lakes of Wisconsin since 1897, have been able to study the quantity and nature of the soluble matter in fresh water over the past few years with quite refined methods. First, we mention the study by Domogalla and Fred (44) on ammonia and nitrate.

In this study, ammonia was at a maximum in May (550 mg/m^3) and at a minimum in August (50–200 mg/m^3), while in December a second maximum was reached (400 mg/m^3). The nitrification process (NH_3 oxidation to NO_3^-) followed the ammonia levels, with the most powerful nitrification taking place near the NH_3 maximum. Nitrate was also at a maximum concentration in spring (April 175 mg/m^3) and at a minimum right after summer (Aug. to Sept., 50–100 mg/m^3). The bacterial denitrification process (NO_3^- reduction to N_2) appeared to be at a maximum in spring and right after summer.

The work done by Birge and Juday (28) on the levels of organic matter in lake water is also very important. In 1909, Pütter (103) posed the hypothesis that many organisms obtain their food not from other organisms or detritus, but from dissolved organic matter. This hypothesis is regarded as controversial without ever actually having been tested. From the above-mentioned research by Birge and Juday it seems that the quantity of dissolved organic matter can considerably exceed the quantity of plankton.

Lake Mendota waters have an average concentration of organic matter of approximately 15 mg/L. Of this, almost 2 mg/L is plankton; the rest is in

---

[5]Hard water has high concentrations of the sum of Ca^{2+} and Mg^{2+}.

(mainly colloid) solution. This solution contains e.g., proteins, lipids, and carbohydrates. Birge and Juday state, following Pütter's hypothesis: "the question of the fundamental food supply of the lake must be reexamined in view of the presence of these relatively large amounts of dissolved organic matter."

Even if we consider 1 mg/L of plankton as the total harvest of a lake, 1 ha over a depth of one meter will still result in 10 kg of organic matter. However, if we take the total amount of organic matter within the same volume we come to an organic matter amount of 150 kg per year. This is comparable to the [typical] harvest of 1 ha in grain (500 kg) or hay (200 kg). One can say that the productivity of a freshwater lake is equal to that of a farm field or meadow.

Domogalla, Juday, and Peterson (43) have taken a close look at the various forms in which nitrogen is present in a lake. Besides nitrite, nitrate, and ammonium, nitrogen also occurs as protein, amino acids, purine, amines, and amides. These "higher" forms of nitrogen seem to occur most abundantly in the winter. The following table provides an overview of the various forms of nitrogen at the surface and at the bottom of the lake.

Forms of nitrogen in Lake Mendota in mg/m³ (from Domogalla, Juday, and Peterson).		
	Surface, 18/06/1924	**Bottom, 25/06/1924**
Plankton	92.4	44.9
Soluble	515.6	766.9
NH_3	96.0	280.0
Loosely bound NH_3	16.0	20.0
NO_2^-	10.0	17.0
NO_3^-	69.4	92.6
free amino-N	54.0	81.0
peptides	135.0	140.0
$\frac{1}{2}$ tryptophan	5.3	7.0
$\frac{3}{4}$ arginine	31.1	34.7
$\frac{2}{3}$ histidine	3.8	6.8
Amide-N	12.4	19.3
Purine-N	8.4	9.5
Amine-N	14.2	16.0
Total soluble	455.6	724.3
Not determined	60.0	42.6

This is a very comprehensive table, and the authors do not hesitate to explain that these compounds most likely play an important role in the biological cycle.

We consider fresh water a rich environment, especially rich in the so-called limiting compounds. No wonder that the development of microorganisms can

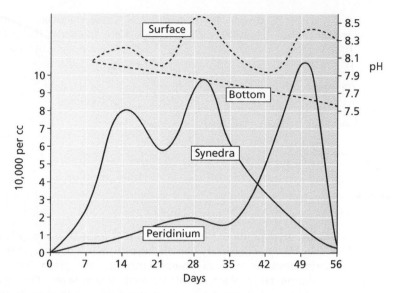

Figure VIII.3 Influence of organisms on the pH level of fresh water.

sometimes occur with explosive fierceness, until a limiting compound (e.g., phosphate) is depleted, or until temperature or light becomes a limiting factor. Organisms such as diatoms or dinoflagellates sometimes occur with a concentration of over 30 million individuals per liter. Such an epidemic, which occurred in Searsville Lake, California, in 1924, is graphically displayed in Figure VIII.3. First the diatom *Synedra pulchella* appeared, reaching its first maximum after two weeks and its second maximum after four weeks, after which it quickly decreased in abundance. The dinoflagellate *Peridinium*, which was already present in lesser quantities, experienced a short but powerful period of growth. Both organisms are photosynthetic, and the pH level of the top layer of water [epilimnion] (dashed line in the figure) fluctuated up and down along with the development of the organisms, while the pH level of the water layer at the lake bottom [hypoliminion] decreased (perhaps due to the enhanced bacterial effect on the detritus).[6]

Examples such as these can be found in all limnological studies. Based on the above-mentioned data, and also in accordance with what was said in the general chapters on the environment, we can now design a general picture of a eutrophic freshwater body. Figure VIII.4 shows a cross section (not to scale) of a reservoir lake. The cross section of the dam is drawn on the left, with depth given in feet. The thermocline, or transitional layer, which separates the

---

[6]Presumably Baas Becking is referring here to the microbial decomposition of the organic detritus.

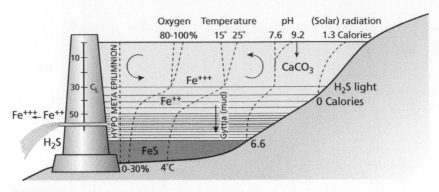

Figure VIII.4

epilimnion from the hypolimnion (see Chapter V), is located at a depth of about 30 feet. The factors necessary for the synthesis of organic matter (through photosynthesis as well as chemosynthesis) are given for the epilimnion. Circulation takes place in this aerated, warm layer. The light intensity is optimal here, photosynthesis occurs along with an increase in the pH level of the surrounding environment (Chapter V), and calcium carbonate sometimes precipitates here (*Chara, Cladophora*). There is a large development of plankton here, which, after dying off, sink to the bottom as a rain of detritus. Due to the high oxygen content, iron (colloid) is present as ferric iron. Light and temperature (especially the former) are the most important physical factors limiting the development of aquatic weeds, such as e.g., W. Beyerinck has proven for the bog waters in the province of Drenthe (26). Of the chemical limiting factors, we have already introduced phosphate, which accumulates in the depths of the lake and only becomes available again when the lake waters mix in spring and fall (in a deep lake) or when the materials at the lake bottom are carried to the surface by vertical flow (in a shallow lake). Dead organisms, calcium carbonate, fine particles of clay and sand, etc., will slowly sink into the hypolimnion and produce mud or *gyttja*. On top of this mud lies a completely different biotope: there is little [or even no] oxygen and the temperature is low. Decomposition processes predominate, and large amounts of carbonic acid and organic acids are produced. Iron is found here in ferrous form and partially bound as sulfides. If the overlying water layer is not too thick and the amount of light is not too greatly reduced, the H_2S will be processed by anaerobic colored bacteria.[7] The formation of methane from cellulose, the process of denitrification, the formation of hydrogen, and even traces of CO occur in this zone. The hypolimnion is a destructive layer where oxygen is used, and it is located under the epilimnion, a synthetic,

---

[7]Baas Becking is referring here to the oxidation of the H_2S by anoxygenic photosynthetic bacteria; these bacteria play prominently in some of the other chapters.

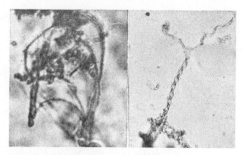

Figure VIII.5 Iron bacteria, greatly enlarged/magnified. Left: *Toxothrix*. Right: *Gallionella*. (Photo by L. Daugherty.)

oxygen-producing layer. In this way, the thermal stratification of water can be enhanced by biochemical stratification (Thienemann, 117).[8]

In Figure VIII.4 the lake is depicted as a reservoir. This is done in order to include an interesting part of the cycle, namely the oxidation of ferrous to ferric iron. Although Searsville Lake in California is not deep, there is a dam from which deep water can be discharged.

At the locations where the discharges are taking place, the ferrous iron (e.g., occurring as bicarbonate) oxidizes to iron (III) oxide-hydroxide:

$$4Fe(HCO_3)_2 + O_2 + 2H_2O = 4Fe(OH)_3 + 8CO_2$$

whereby the acidity of the water increases. Here, the process shown as a lab experiment in Chapter V occurs naturally. In this location, and in general everywhere where poorly aerated, iron-rich water is exposed to the atmosphere, iron bacteria can develop. Seeing as how these forms seem to grow especially at lower temperatures[9] (with the exception of *Leptothrix ochracea*), aerated deep water is a hospitable environment.[10] In our country one can observe them especially in wintertime anywhere where ferrous iron comes to the surface. The colonies fix themselves as large orange masses to all kinds of objects in the water while a thin, hard, often ragged iridescent film of Fe_2O_3 floats on the water. Some forms of these bacteria are shown in Figure VIII.5. The one on the left belongs to the genus *Toxothrix*; that on the right, twisted like a bobby pin, belongs to the genus *Gallionella*. The long threads are encrusted with iron (III) oxide-hydroxide. Cholodny (35) discovered at the end of the thread an organism which secretes the whole thread of slime, covered later

---

[8]I don't think Baas Becking means that the heat produced by microbial metabolisms adds to the thermal stratification, as seemingly implied, but rather, that the thermal stratification inhibits mixing to an extent, allowing chemical stratification.
[9]Not clear where this idea originates, but it has not been retained in modern thinking about the distribution of Fe-oxidizing bacteria.
[10]Presumably Baas Becking is referring to the interface between aerated (colder) deep water and ferrous Fe accumulating in sediments or deep depressions in the lake bottom.

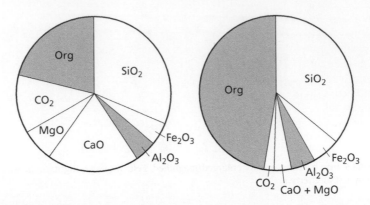

Figure VIII.6 Phosphate level of the surface water in the Channel, as well as hours of sunlight. Both as a function of time. From Harvey.

with hydroxide. According to some (Perfiliev, 96), these bacteria play a major role in the formation of various iron ores.[11]

Not only the ferrous iron, but also other chemical species can be oxidized when brought from the bottom water layer to the oxygen-rich top layer. This does not always happen spontaneously as with Fe^{2+} (and possibly also partially with nitrite) but occurs under the influence of bacteria that process compounds such as NH_3, CH_4, H_2, and CO.

Bog water, as described in the previous chapter, is in fact entirely a hypolimnion. The constructive processes take place in the growing peat moss just above the water table, and the submerged peat is not able to keep the oxygen pressure stable. Indeed, in eutrophic water decomposition can predominate. If the H_2S concentration reaches levels higher than 10^{-3} molar, most algae will die off and the whole pond or ditch can become an ink-black mass (a description of this process can be found in Jordan, 74). Theoretically, if the destructive metabolism equaled the synthetic one, the organic products produced would be completely metabolized. However, there is always a large amount of organic matter captured in the bottom mud. Schuette and Alder (110) report that the pondweed *Chara* is able to capture 427 tons of carbon in the form of carbonate in one year in a single lake (Green Lake, Wisconsin) (993 tons of calcium carbonate). Regarding the nature of the deposited matter, work done by Black (29) has provided more details. Figure VIII.6 provides a schematic overview of the soil deposits in a typical "hard water lake" (left) and in an oligotrophic lake (right). In the latter, organic matter can reach up to 50% of the dry weight of mud, while apart from the kieselgur [diatomaceous earth], the mineral content occurs in small quantities (less than 15% of total dry weight). A typical decomposing mud from eutrophic water (saprophytic) contains a lot of Ca, Mg, carbonate, kieselgur, and organic matter. The analyses by Black show that

---

[11]This idea has been resurrected in modern thinking about the formation of banded iron formations.

much of the matter that is synthesized at the surface can be stored in the mud at the bottom. Raised bog would be an extreme case, as organic matter would predominate here. Roughly speaking, the percentage of organic matter in the mud at the bottom of a lake is inversely proportional to the intensity of the bacterial activity which takes place within the lake. The most intense bacterial life can be found in eutrophic water. In acidic mud in soft oligotrophic waters, processes take place which lead to the formation of peat and lignite. Carbonization [and the formation of peat and lignite] does not take place in alkaline to weakly acidic saprobic eutrophic water or seawater.[12]

EDITOR'S NOTES

Baas Becking has managed to capture the geobiological essence of limnology in a single chapter. As in many of the previous chapters, he begins by defining the chemical environment and proceeds to discuss the interplay between the chemical environment and life. In particular, he shows us how biological processes – sulfate reduction, for example – can influence the major ion composition of productive lake waters. The influence is direct though the process itself (the removal of sulfate and the production of bicarbonate) and indirect through the influence of sulfate reduction on the pH of lake waters and the subsequent removal of calcium and bicarbonate through calcium carbonate precipitation. Baas Becking also describes how photosynthesis influences calcium carbonate saturation and precipitation.

Of particular importance is the discussion of how the limiting nutrients nitrogen and phosphorus control lake productivity singly and together, and how this control varies seasonally with light, temperature, and the chemical stratification of the lake. He also illuminates, in a very modern way, how nitrogen is not only a limiting nutrient, but is also involved in biological transformations including denitrification, and that these nitrogen transforming processes also have seasonal regulation.

One can see that at the time of writing, limnology was a rather young field and Baas Becking, as elsewhere in the book, draws on a combination of the evolving literature, but also on personal observations and the observations of his friends and colleagues. One is struck by how prescient some of these observations are. One example involves the chemical nature of stratified anoxic lakes and, in particular, the accumulation of Fe^{2+} in the anoxic waters and the roles of microbes in oxidizing the Fe^{2+} in the presence of oxygen. Many of these points still drive the literature today. We, of course, know much more about the details of these processes, and the microbial populations and processes found in stratified eutrophic lakes. However, what we know today rests squarely on the foundation presented for us by Baas Becking in this chapter.

---

[12]This paragraph in general would not represent the modern view. While we would agree that bacterial activity would be much more active in eutrophic environments, these environments would also generally be viewed as more organically rich. The raised peat bog that Baas Becking uses as model for oligotrophic environments is not generally representative of oligotrophic environments, which are mostly very organic-matter poor.

CHAPTER IX
Oceans

Nowhere else where life is possible, probably in no other place in the Universe except another ocean, are so many conditions so stable and so enduring.

This motto, derived from the work by L. Henderson entitled *The Fitness of the Environment* (59), contains a grain of truth but can be misleading in its generality. Nonetheless, in this chapter we can safely adhere to it, as only the physical–chemical environment will be discussed. The stability should not be equated with uniformity, however, because in doing so one might be inclined to turn the motto around. In no other aqueous environment does one find a greater variety of biotopes – from the littoral (coastal) zone, with its richness in nutrients, vertical currents, and great variety in light and temperature conditions, to the pelagic zone (open water), with its thin "synthetic layer," and the deep with its own remarkable living conditions. It is precisely in the ocean that the "topos," the location, becomes important, and it is here that we see the "adapted" organisms in their own niche. The most studied marine organisms are the so-called higher organisms. They didn't arrive in their topos through the principle that "the environment selects" or because they are "latently present"; they have, shall we say, selected their environment themselves and have adapted to it.

Yet due to its great mass and the active mixing which takes place within it, the chemical and physical properties of the ocean environment are both stable and sustainable. This environment is also stable because it consists entirely of water (see Chapter V).

Seawater is characterized by its salinity level, which is so high that it considerably influences the physical constants of the solution. Examples are the freezing point, which lies around $-2.3\ °C$, the density, which is 1.026 at 15 °C, etc. Salinity also controls other properties of seawater. For instance, seawater becomes thermally stratified despite the absence of a maximum density at $4\ °C$,[1] thanks to the two influences on density (temperature and salinity). The influence of salinity on the permeability of (solar) radiation was referred to in Chapter III.

---

[1]As is true in fresh water. For seawater the maximum density is reached at the freezing point.

Baas Becking's: Geobiology, Or Introduction to Environmental Science, First Edition. Edited by Don E. Canfield.
© 2016 John Wiley & Sons, Ltd. Published 2016 by John Wiley & Sons, Ltd.

Seawater is a solution which can contain a maximum of 41,000 mg/L of salts (Mediterranean Sea);[2] the average is 34,500 mg/L. In areas where it mixes with fresh water, a range of [salinity] gradients is apparent (Ringer, 106; Massink, 81). In 1884 Dittmar provided his famous seawater analyses, collected on the Challenger expedition.[3] The averages of his 77 analyses are portrayed in the following table:

Substance	Percent	Percentage of total
Na	1.049	30.59
Mg	0.130	3.79
Ca	0.041	1.20
K	0.038	1.11
Cl	1.896	55.27
SO_4	0.263	7.66
CO_3	0.007	00.21
Br	0.0066	0.19

A large number of other substances also occur in seawater. An overview of these can be found in the works of Harvey (58) and Clarke (36).

The substances which occur in seawater can roughly be divided into two groups: those which undergo quantitative change through biological influences, and substances which undergo this change to a lesser degree. In the first group we place I, Br, CO_3, O_2, NH_4, NO_3, PO_4, and SiO_2; in the second group we find Na, Cl, SO_4, and N_2. Transitional between the two groups we find Ca and Mg. This does not mean that living organisms exert no influence on the concentration of sodium or chloride; however, the quantity of these compounds is so large that the changes are not noticeable. Seawater also contains such a large amount of sulfate that it is impossible to detect any changes due to sulfate reduction;[4] these can be observed in other bodies of water but not in oceans. The nitrogen bond is most likely dependent upon the local concentration of carbohydrates or higher alcohols, but is insufficiently studied (as are most bacterial processes in the ocean), such that the ranking of nitrogen with matter not influenced by organisms is quite arbitrary.[5]

---

[2]This would be the maximum in an open ocean setting. In restricted settings, evaporated seawater can hold much more salt than this.

[3]We sometimes take for granted that the major ion composition of seawater has always been well known, but Baas Becking reminds us that this is not the case.

[4]Sulfate is well mixed in the oceans, so there are no real biologically induced gradients in sulfate concentration under normal conditions. However, gradients can be observed under anoxic conditions, particularly in sediments when sulfate reduction is active and where the sulfate supply is limited by diffusion. We also know that while sulfate is well mixed in the oceans, the sulfate concentration of seawater is very much influenced by the intensity of sulfate removal by sulfate reduction.

[5]Indeed, N_2 in the oceans is influenced by both nitrogen fixation and N_2 formation from denitrification and anammox ($NO_2^- + NH_4^+ \rightarrow N_2 + 2H_2O$). Careful measurements now show that the concentrations of N_2 in seawater are affected by these biological processes.

The *variable substances*[6] are involved in the vital [biological] cycle. Theoretically one can predict the entire cycle and identify the presence of various compounds through enrichment cultures. However, this has not clarified the role these (mainly bacterial) compounds play in the ocean. A thorough study of "models," in this case seawater aquaria, should provide us with a large amount of fundamental data (Honing, 64). Especially with regard to the nitrogen cycle much remains unknown. It seems that nitrification does occur, but little is known about this. The nitrate level, which in deeper water can be quite high (200 mg/m³), can become a limiting factor or even disappear completely in surface water (up to 5 m) in late summer, as Harvey was able to point out for the English Channel. At the same time, the water at the [sea] floor contained important amounts of nitrate. As denitrification should also occur in anaerobic circumstances, the main cause of the disappearance of nitrate in surface water must be the phytoplankton. If they are absent or extremely repressed, the nitrate can accumulate (seawater aquaria).

While the level of ammonia can reach over 100 mg/m³,[7] active nitrification will reduce this level. In light, especially when there is much phytoplankton, it appears the NH_4 level increases. The amount of soluble organic matter in the littoral zone can equal that of fresh water. A summer maximum [in dissolved organics] has been noted here; however, a systematic chemical analysis, as provided for fresh water by the Wisconsin Survey, is still lacking. The level of silica in seawater (approx. 200 mg/m³) is highly influenced by silica-absorbing organisms (diatoms). After a great bloom of these organisms the level of SiO_2 dropped to 20% of the original level, without completely disappearing. In 1925, the author found the marine diatom *Aulacodiscus kittoni* Arnott in huge quantities just past the mouth of the Columbia River in the state of Washington. In 1928, in a river mouth near Corinto, Nicaragua, great numbers were found. It is possible that the increased level of SiO_2 in the water where it is subject to mixing, as well as phosphate and nitrate, plays a role, while the osmotic and antagonistic factors should not be ignored. Nitrate and phosphate levels are limiting factors "par excellence." Our knowledge regarding these factors is due mainly to the work of the Plymouth Biological Station. Figure IX.1, derived from Harvey (58), shows the phosphate levels of surface water in the English Channel 20 miles southwest of Plymouth over the course of a year, in relation to the number of hours of sunlight. The inverse relation of sunlight to phosphate level is obvious. The plankton which develops with light seems to be able to rob the water completely of phosphate.

Regarding the level of iodine, varying levels are found in the literature, of which the largest range cannot be certified because of mistakes in the methods. The largest amount is close to 3 mg/L. It is possible that the variability can be explained by the update of the element by sponges, brown algae, and red algae.[8]

---

[6]That is, substances with variable concentrations in seawater affected by biological processes.
[7]Such high concentrations are not typical of ocean water and are only found in some anorie basins and fjords and in anoxic marine sediments.
[8]It is now known that iodate is the stable form in seawater, showing modest surface-water depletions due to uptake by algae.

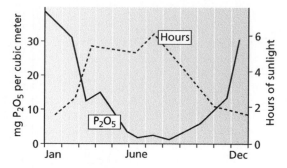

Figure IX.1 Phosphate level of the surface water in the Channel, and hours of sunlight, both as a function of time. (From Harvey.)

Experiments on this are needed. Also bromide, although amply present, can sometimes completely disappear.[9] For this as well, it is unknown whether organisms play a role.

When "organic" matter or a limiting factor ends up in the bottom sludge, the substance can become lost to the cycle, at least temporarily. Upward flow can make this matter available again. This is not the place to discuss the causes of the circulation of seawater. Yet this circulation is one of the most important processes, as the cycle in the ocean is broken if circulation stops. Oxygen can then act as a limiting factor. This gas is less soluble in seawater than in fresh water (the difference is approximately 20%). In the whole range of oxygen pressure, there are certain regions to which particular organisms have adapted. When the pressure is too high, this can be harmful to many fish species, although on the other hand many mud-dwelling organisms, equipped with oxygen-binding pigments (hemoglobin, hemocyanin), are able to live even at very low levels of oxygen pressure. The sulfur cycle in the ocean is caused by the same organisms as in fresh water. Iron bacteria can also be found in the ocean.

Of the heterotrophs, it is mainly the bacteria that are able to liquify agar, and the bacteria that eat away chitin, which play specific roles in the marine environment, given the large amount of algae and crustacean material which is decomposed by bacteria. The study of these forms has only just begun.[10]

The most important biogenic chemical factor is carbon dioxide. This gas is bound in the upper layers by phytoplankton. Atkins calculated that the English Channel contains 140 tons of organic matter, in the form of plankton, per km². In Chapter V it was pointed out that the pH of water can increase through the assimilation of carbon dioxide.

---

[9]This must be due to analytical uncertainties, as bromide is well mixed in seawater and its concentration is only influenced by evaporation and dilution.

[10]At the time of writing only a relatively few types of marine heterotrophic bacteria had been enriched. It now known that the diversity of marine heterotrophs is virtually countless.

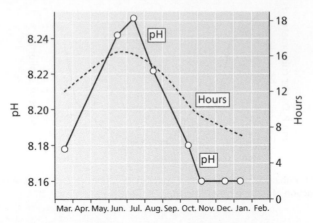

Figure IX.2 pH and hours of sunlight as a function of season. (From Harvey.)

As originally pointed out by Palitsch, seawater has a very constant pH (see also Moravek, 88), namely 8.1–8.3.[11] Atkins has been able to show that during the summer the alkalinity of the surface water barely increases, as can be seen in Figure IX.2, where pH and hours of sunlight are given as a function of season. In shallow pools this phenomenon can be more pronounced.

When considering the so-called ionic balance of seawater (i.e., the amount of cation and anion equivalents, see Figure IX.3) we can observe that there is a variable part, indicated by CO_3 (including the other forms of carbonic acid). In Chapter V it was deduced that the maximum level of bicarbonate is reached at a pH of 8.2. This means that seawater carries a maximum of bicarbonate ions. Furthermore, seawater has a fairly constant amount of so-called base excess, meaning that the quantity of cations is in equilibrium with H^+, OH^-, CO_3^{2-}, and HCO_3^-; namely 23–26 × 10^{-4} of normal,[12] which can be determined through titration with 0.01N HCl to an end point of pH = 4 (methyl orange). If the base excess and the pH of seawater are known, then with the aid of the Johnston equation (see Chapter V) the amount of carbon dioxide can be determined. It seems that seawater is close to equilibrium with the amount of carbon dioxide in the atmosphere. At a pH of 8.3, seawater absorbs carbon dioxide; when the pH is less than 8.2 it emits CO_2. McClendon constructed a well-known nomogram (Figure IX.4) from which one can graphically determine from the pH and base excess the total amount of carbon dioxide. It appears that seawater contains CO_2 in amounts of 40–55 cc/L[13] in either free or bound form.

---

[11]This would be surface seawater. The values are slightly lower now due to the accumulation of anthropogenic CO_2 in the atmosphere. At depth in the oceans, the pH of seawater can be reduced to values of 7.6–7.7.

[12]This is quite close to the molar concentration of HCO_3^- in the oceans.

[13]This is equivalent to the number of cubic centimeters of CO_2 gas per liter, at standard temperature and pressure. Thankfully, these units are no longer used to express concentration.

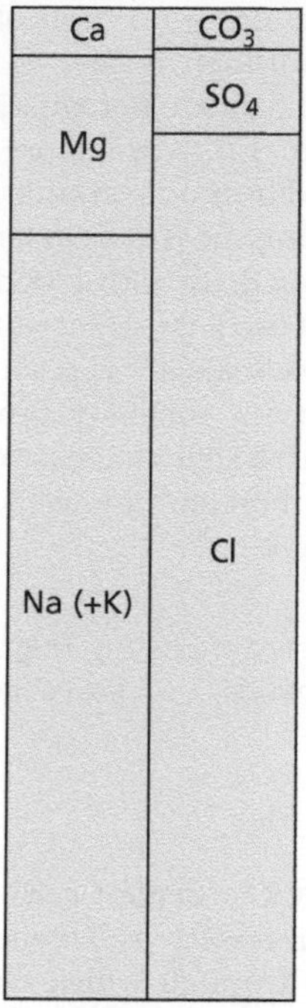

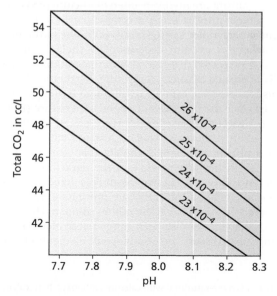

Figure IX.3 Ionic balance of seawater. (From data by Dittmar, 42).

Figure IX.4 The relationship between amount of carbon dioxide, base excess, and pH in seawater. (From McClendon.)

A few things are of course also a function of temperature, pressure, and salinity, such that a more thorough review of the carbon dioxide issue is required. This carries quite some difficulties with it, which lie outside the scope of this work (though see Buch, 32). A few more remarks must be made here regarding carbon dioxide equilibrium. Seawater is buffered through the base excess, which is able to neutralize the H-ions in large quantities without changing the pH. One can use an analogy with blood and use the term "alkaline reserve." The analogy does not carry far, as seawater is poorly buffered towards the alkaline (pH 8.2–9.2). Also, seawater is practically saturated with calcium carbonate.[14] This substance has a solubility product (calcite) of 9.8×10^{-9}. $CaCO_3$ already precipitates with a slight increase of the CO_3 concentration. For the precipitation of calcium carbonate, various biogenic causes can be given, of which a few are mentioned here.

	Secretion of base (e.g., NH_3)		A
Increase of pH by	Absorption of CO_2 from water	Chemosynthesis	B
		Photosynthesis	C
		Other causes (?)	D
Emission of $CaCO_3$ by cells (?).			E

A specific bacterium needed to deposit calcium, as Drew (46) claimed to have discovered, is in fact unnecessary, as Bavendamm (20) has proven. There are many processes involved through which either a base is excreted (e.g., *Ulva*) or CO_2 is absorbed.

In both cases it [calcium carbonate precipitation] results either in encrustations or, if microorganisms are involved, often in calcium [carbonate] sediment. An increase in temperature is often accompanied by a change in the CO_2 tension (calcium sinters at hot springs). However, one of the most mysterious processes in seawater is calcium deposition by shells, sponges, corals, and coralline algae. It is suspected that the calcium (magnesium) is absorbed by the cells, and locally deposited as calcite (aragonite, magnetite, dolomite). It is not known how this process came into being, although Irving and the author have found that for red algae the process is of a combined nature (Figure IX.5). Five grams of the red algae *Corallina officinalis* were placed in 200 cc of seawater bubbled with air. This setup was regularly titrated with 0.01N hydrochloric acid. In 40 hours of continuous light the algae appeared to have practically depleted the $CaCO_3$ in the water, while in the dark this process took much longer. The difference between the two treatments demonstrates the influence of carbon dioxide assimilation [on carbonate precipitation].

It has been noted that seawater is sometimes compared to blood with regard to its buffering capacity. It has also been compared to blood in a different context, which brings us to discuss the non-biogenic elements of seawater.

---

[14]We now know that it is over-saturated with calcium carbonate, both calcite and aragonite. These minerals do not spontaneously precipitate, for kinetic reasons.

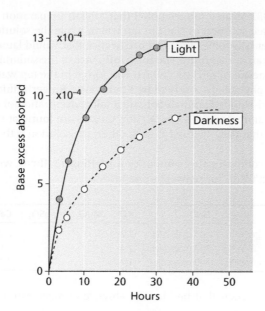

Figure IX.5 Base absorbed by coralline algae in light and in darkness. (From Irving & Baas Becking.)

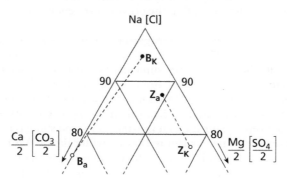

Figure IX.6 Triangle diagram (upper part) in which is depicted the anion ratio of human blood B_a and seawater Z_a; the cation ratio of blood B_k and seawater Z_k; all in equivalents (orig.).[15]

In Figure IX.6 the points Za and Zk represent the equivalent ratio of anions and cations in seawater, in the upper part of the triangle diagram. The ratio in which sodium occurs in relation to the amount of calcium is the same as has been found for blood (Bk in the figure).

---

[15]"orig." indicates a figure that is original to Baas Becking's work, not based on the data or design of others.

When in 1882 Sidney Ringer noted that a heart preparation continued to pump in a diluted solution of NaCl in tap water, while the solution of this salt in distilled water was toxic, it was an occurrence that would later be described by Loeb (79) as an antagonism. The toxic influence of the sodium was neutralized by the influence of calcium [which was found in the tap water].

The animal physiology is served by fluids called Ringer's fluids, which, for example, in cold-blooded animals contain a salt concentration of 0.65% and in warm-blooded animals 0.95%. Na, K, and Ca salts are found at such a ratio in these solutions that they "detoxify" each other; the solution is then considered "balanced."

When one compares the composition of Ringer's fluids with a formula provided by van't Hoff for seawater, one finds:

	NaCl	KCl	MgCl$_2$	MgSO$_4$	CaCl$_2$
Seawater	100 molecules	2.2	7.8	3.8	2.3
Ringer 0.65%	100 molecules	2.4			1.6
Ringer 0.95%	100 molecules	1.7			1.1

Here it can be seen that besides the absence of magnesium salts, there is some similarity.

Höber takes this idea quite far by stating: "The ratio of some ion concentrations in the Ringer's fluids is similar to that of the ions present in seawater. This means that one can almost say that the organs of vertebrates are best preserved by some kind of seawater."(61)

The total absence of sulfate and magnesium in the Ringer's fluids, as well as the [salt] concentration, which is little more than a quarter of that of seawater, illustrates the fact that there is little similarity between Ringer's fluids and seawater. Yet there is still a reason for Höber's statement, which can be found in the theory, originally posed by Macallum (80), according to whom blood can be seen as Paleozoic seawater, which, caught in the circulatory system, survived the jump from aquatic life to terrestrial life. Blood still has characteristics similar to those in seawater, as Figure IX.6 shows for the ratio Na : Ca. Yet one must not look much beyond this, as further review of this figure indicates.

Seawater is rich in sulfate, while blood is poor in it. The carbonate level of blood is very high. Any hydrologist, shown the ionic composition of blood serum, would most certainly not think of seawater, but much sooner of a chlorocarbonate solution as occurs in deep groundwater or in deserts (Chapter X).

People have also tried to find the evolutionary trail in comparative physiology. It indeed appears that bodily fluids of lower life forms (Echinoderms, Crustacea, Cephalopods) contain cryoscopic qualities which are similar to the surrounding seawater, enabling equilibrium between the internal and external environment in accordance with the classic osmotic theory. Furthermore, to a certain extent the osmotic pressure of the internal fluids follows that of the external environment.

The same was discovered for rays and sharks – cartilaginous fish – while bony fish were able to regulate their own osmotic pressure, resulting in their blood's freezing point depression shifting from 2–3 °C to 0.7–1 °C. The same is true for the sea turtle. It now seems that animals with their "own" osmotic environment, such as bony fish, have a lower concentrated fluid level, possibly originating from ancient seawater from the time that "the sea was not yet so concentrated." However, there are some facts which oppose this opinion. First, the composition of e.g., shark blood has little to do with seawater; a large part of the freezing point depression here is caused by urea. For crustaceans it is often glucose, formed from glycogen, which controls the osmotic pressure.

Secondly, it is very unlikely that the archaic seas would have had such a strange consistency as displayed in Figure IX.6, a consistency analogous to nothing yet known on Earth. This second point of resemblance between blood and seawater seems to the author to be less well chosen than, for instance, the comparison of the so-called buffer capacity.

The fact of antagonism remains, even though seawater does not seem to be as perfectly universal a solution as Loeb and his school would have us believe. From earlier works it was already clear that calcium and sodium are able to neutralize each other's toxicity. This occurred in the ratio Ca : Na = (1–2) : 100.

Furthermore, several rules regarding antagonisms were established, which, inspired by the characteristics of seawater, are incorrect in their generality. The first rule was that the ratio Na/Ca, or rather (Na + K)/(Mg + Ca), which is most suitable for this antagonism, would be approximately equal to the corresponding ratio in seawater and that the most suitable ratios would be independent of the concentration. In the chapters on fresh water and brine one can read about completely different types of antagonisms. It is likely that in many cases in fresh water the Na/Ca ratio is smaller, and in highly concentrated brines much higher than 50, the value that Loeb gave. People have tried to prove that wheat grows "best" in diluted, artificial seawater. The mistake made here is that only some of the ion combinations have been tested and not, as described in Chapter X, the whole range of possible combinations. One will never be able to describe the antagonistic state with results from only a few experiments.

It has also become apparent that antagonisms occur not only between mono- and bivalent cations but also between the most variable cations and anions.

Some examples will be discussed in the next chapter.

EDITOR'S NOTES

As in the other chapters, Baas Becking seeks to place the organism in the context of the environment, in this case the oceans. In this chapter he makes some spot-on observations about the role of limiting nutrients in controlling primary production and relates some interesting findings, including some of his own, on the process of carbonate precipitation. He devotes most of the chapter, however, to exploring the relationship between the internal fluids of organisms and seawater, and the physiological and evolutionary significance of this.

Overall, however, one is struck by how, in Baas Becking's time, so little was known about the "geobiology" of the seas, including its microbial ecology. At the time of writing oceanography was an emerging field and little was known about the microbial ecology of seawater. The sulfidic waters of the Black Sea and the Cariaco Basin were yet to be discovered, and the analytical techniques needed to explore the chemistry of seawater were still evolving. Indeed, our understanding of the microbial diversity and ecology of seawater has expanded enormously in the last 15 years with the advancement and clever application of molecular tools. None of this was available in Baas Becking's time, where, much more than today, the oceans were a vast unknown frontier.

CHAPTER X

Brine

By "brine" we mean saline solutions with concentrations higher than those of river water, excluding seawater and its dilutions. Brine, which in a saturated state can contain over 400,000 parts per million of soluble particles, is the opposite of the electrolyte-poor raised-bog waters (30 parts per million).

1 Aeolian brine. We have come to know that the typical raised bog is rain-fed. Raised bogs can only exist in climates where there is more precipitation than evaporation. The electrolyte in this raised bog water is carried in by ocean winds. In a dry region where sea salt is transported by dry desert winds, this can accumulate as "aeolic salt." This is the case with Lake Sambahar in northern Rajputana, India, which receives its salt solely from the dry western wind.

According to calculations, approximately 3000 tons of solid matter is deposited in this lake annually, mainly sodium chloride and sodium sulfate.

2 Connate brine. We will briefly mention here the types of brine that originate from solutions from older salt deposits. As these layers are often quite specific and complex, the characteristics of these types of brine are often quite remarkable (e.g., the Dead Sea with its high levels of calcium and bromide).

Of greater importance to us are the types of brine which are formed by the leaching of rocks. In these cases, the nature of the rocks determines the composition of the solution, meaning that the composition of the solution can already be predicted at the beginning of the leaching process, based on the composition of the headwaters.

3 The most common case is river water – especially where the water runs through sedimentary rock. Its composition can be displayed in the familiar triangle diagram (Figure X.1). A_1 depicts the ratio of anions, K_2 that of cations of the river water. Through the meandering of the river bed it seems that the main exchange is that of calcium and carbonate (and bicarbonate) for sodium and chloride. The end solution, at around 30,000 parts per million, has approximately the same composition as seawater (A_2K_2). Also, waters that have never been in contact with the sea or marine deposits can have an analogous composition (Great Salt Lake, Utah). From the diagram one can

Baas Becking's: Geobiology, Or Introduction to Environmental Science, First Edition. Edited by Don E. Canfield.
© 2016 John Wiley & Sons, Ltd. Published 2016 by John Wiley & Sons, Ltd.

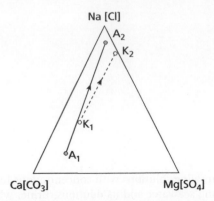

Figure X.1 Diagram of river water and seawater. (Data from Clarke.)

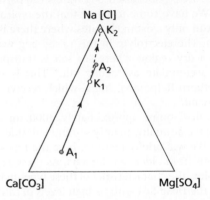

Figure X.2 Diagram of the composition of the Truckee River and Lake Pyramid.

see that the water A_2K_2 is characterized by low concentrations of calcium and carbonate. As with seawater, the pH level is never high (pH = 8–9).[1]

4 Waters flowing through magmatic rock generally have different traits. A good example of such a system can be found in the waters of the eastern slopes of the Sierra Nevada in the states of California and Nevada. Here a mountain lake (Tahoe) flows into a river (the Truckee), which ends in the desert in two lakes: Pyramid and Winnemucca. The composition of the Truckee river water is shown in Figure X.2 as A_1K_1. While the ratio of anions is comparable with that in Figure X.1, it can be noted that the location of the cation point K_1 is very different; the amount of calcium is much less than that of other cations. In Pyramid Lake, where the total [salt] concentration is only 0.1%, this composition of the water results in practically all the calcium and magnesium being deposited as carbonate.

---

[1]Actually, a pH of 9 is pretty high for a natural water.

Figure X.3 Algal tuffs on the shores of Pyramid Lake, California. (Photo by the author.)

The composition of the water in Pyramid Lake is given by the points A_2K_2. The precipitation of these carbonates can be viewed as a biological process in so far as Cyanobacteria (*Aphanocapsa*) play a central role in the formation of the oolites and tuffs that define these lakes. The erratic, often pyramid-shaped masses of these tuffs, which can best be seen after the water level has receded (which must have also occurred in historical times), is seen in the photo (Figure X.3). Similar [ion] concentrations occur in nearby Winnemucca Lake, as well as in Walker Lake and Humboldt Lake.

With this process [the precipitation of carbonate], whereby it is likely that the photosynthesis of the cyanobacteria plays a role, almost all of the calcium and magnesium is removed from the water, and the more concentrated solutions which belong to this or an analogous system (Sand Springs, Nevada; Soda Lake, Nevada; Owens Lake, California; Searles Lake, California) have a consistency which differs only minutely (i.e., in the levels of carbonate and sulphate) from those found in Pyramid Lake, as depicted by the points A_2 K_2 in Figure X.2.

From the location of point A_2, it follows that even when Ca and Mg are removed, a large quantity of carbonate (and bicarbonate) remains. The final brine solutions consequently are very alkaline (pH = 10–11).

5 Let us consider seawater (A_2 K_2 in Figure X.1), which in warmer countries is evaporated in salt pans by solar heat and continues to provide a large percentage of our table salt. This seawater, which contains salt at a molar

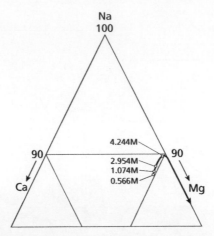

Figure X.4 Diagram of the shift in the ratio of cations during the evaporation of seawater. (From data by Usiglio, according to Jacobi and Baas Becking.)

concentration of approximately $0.55,^2$ will, once this reaches 0.90, have already lost the calcium carbonate. The brine will nonetheless contain a large quantity of calcium that is the first to precipitate as a sulfate [gypsum, $CaSO_4 \cdot 2H_2O$] at much higher concentrations (2.7 molar NaCl or 17% total salt in the solution). Sodium chloride follows calcium sulfate as the next to precipitate, resulting finally in brine (bittern) that contains mainly chlorides and sulfates of Mg and K. For the cations these changes are displayed in Figure X.4. We will come back to these environmental shifts; it needs only to be mentioned here that the two types of solutions, which we will call "sedimentary" and "magmatic," not only have different compositions from the beginning, but, when concentrated, they also produce brines with a totally different composition.

The analysis of desert brines and brine from seawater is shown in the following table.

Matter in %	Searles Lake	Owens Lake	Salt pan, San Francisco Bay
$MgCl_2$	....	....	8.17
KCl	4.69	3.74	2.57
NaCl	16.63	17.44	17.43
Na_2SO_4	6.98	5.40	5.04
Na_2CO_3	3.46	5.58	....
$NaHCO_3$	0.77	1.43	0.08
$Na_2B_2O_4$	1.39	....	....
$Na_2B_4O_7$	0.43	2.72	0.15
H_2O	65.65	63.69	66.56

---

[2]This is the concentration of Cl⁻ in seawater.

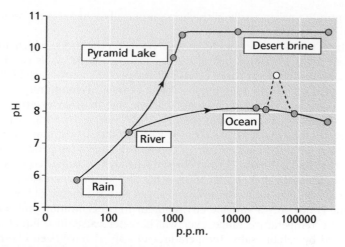

Figure X.5 pH levels of various concentrates.

The great similarity in the compositions of the two desert brines and their large difference from the seawater brine is remarkable.

As was noted earlier, the desert brines are mainly alkaline. Figure X.5 shows the range of pH levels found in the various concentrations of desert brines and in seawater.

Rainwater was used as the starting point in both cases (pH = 5.8). The abscissa shows the concentration in parts per million on a logarithmic scale; the ordinate shows the pH level. The dotted line in the seawater concentrate column shows how the pH level can increase with photosynthesis and can decline after the removal of precipitated calcium. As calcium is not always removed during salt production through evaporation, it is also possible for highly concentrated brines to have higher alkaline levels than shown in this figure.

For further data regarding the composition of natural brines, please see the relevant literature (95,11).

6 **Physical environment.** In general, brines will occur where evaporation exceeds precipitation, i.e., in regions with lots of sunshine (Portuguese coast), with a very dry season (Asian monsoon, Madura salt pans), or in desert-like regions (Gobi, Kalahari, Sahara, etc.). Rains and incoming waters can suddenly reduce the concentration [of salts], although often the layer of fresh water, unmixed with the brine, will remain floating on top. This was observed by the author at the mouth of the Bear River which flows into the Great Salt Lake in Utah from the north. Here, the concentrated brine is covered by a thin layer of fresh water, approximately 10 cm at the source and thinning as it spreads to a layer of a few millimeters. In the salt pans of San Francisco Bay, people cover the brine with a layer of fresh water during the rainy season, as this water does not mix with the brine. An increase in

concentration occurs easily in the brine, especially when it is exposed to the sun in very thin layers.

At high tide, seawater can form small pools on the rocks, which are then exposed for many hours before the sea returns. These pools can become highly concentrated. Nitrogen levels in these pools can also reach high levels due to bird guano, resulting in a particular ecological community.

The composition of brine can also change due to other external factors. Especially in concentrated and saturated solutions, a lowering of the temperature causes the crystallization of various salts. Early one morning in November 1925, when the brine in Owens Lake had a temperature of 2 °C, the author found the lake bottom covered with sodium sulfate crystals, which disappeared again after several hours as the temperature increased. Here, the ionic environment underwent huge changes within a matter of hours.

Brine is not only a variable environment in concentration and composition. The temperature range over which water exists in a liquid state is increased by soluble salts. Temperatures of −30 °C have been observed in desert lakes. Also, when the concentrated solution is covered by a layer of fresh water, the upper layer can build up solar heat such that temperatures of +70 °C have been observed (V. Kaleszinsky, Lake Medve, Hungary). With the creation of this warmer layer, many factors are involved, the main ones being viscosity and density, which through warming result in a "shifting" layer that floats on top of the denser brine but is still denser than the top layer of fresh water.

The specific heat capacity is also reduced as the concentration increases, resulting in concentrated solutions warming up more easily than diluted solutions.

The author has been able to observe this phenomenon repeatedly in the concentrated brines of salt pans. With an air temperature of 19 °C, brine temperatures of 35 °C are not uncommon. This phenomenon does not exist when the solution is covered by a layer of salt. This layer, which in Searles Lake is around 2–5 cm thick, isolates the brine enough that the liquid has a temperature of 22 °C throughout the year.[3]

Table salt is diathermic, meaning it easily lets through infrared radiation. Even in concentrated salt solutions this property cannot be detected, which became apparent from the spectra of NaCl solutions in the infrared range, as Dr. J. Ellis was kind enough to construct for the author.

Even very thin salt layers absorb a large amount of light; the intensity is reduced by the diffraction of the light by the encased air (the crystals are usually white). The top layer of brine in Searles Lake thus receives only 0.01 % of the direct available solar energy,[4] and the organisms which can be found directly under the salt layer are, in terms of light intensity, practically living in the deep sea.

---

[3]This is a very unusual situation where mineral crust forms from evaporating brine on top of the brine.
[4]We have to take Baas Becking's word for it, but this large amount of light attenuation through a 2–5 cm thick mineral crust surprises me very much.

Of the soluble gases, carbonate,[5] especially in alkaline conditions in bound and half-bound states, is present in large quantities. This is different for non-chemically bound gases such as oxygen. With increasing salinity, the oxygen levels decrease, and concentrated brine holds less than one-fifth of the oxygen present in pure water. The organisms that inhabit these liquids, like the organisms that inhabit warm waters, are exposed to low oxygen pressure.

Further analysis regarding the environmental factors of brine lies outside of the boundaries of this dissertation. It suffices to say that this environment can be called extreme on many fronts. For further analysis, please see the literature (11).

In conclusion, we could characterize brine as a solution in which:

The osmotic value can be unusually high,

The ionic ratio deviates from known values for fresh water and seawater,

Temperature is [may be] extreme,

Solar radiation, specific heat capacity, and oxygen pressure are often low,

Viscosity is high,

Toxic substances such as borates and arsenic may be present, and

Furthermore, the above-mentioned characteristics are subject to large and sudden changes.

7 Organisms. Otto Stocker recently wrote (115): "At salt concentrations higher than 10–15%, the halophytes start becoming extinct. Salt lakes with higher concentrations generally appear to be free of vegetation (Dead Sea 19.5%), although some forms of bacteria and algae have been observed at even higher concentrations (halophilic sulfur bacteria up to 22% and *Chlamydomonas* up to 25%)."[6]

This statement gives the general opinion that the stronger brines contain almost no organisms, and that the concentration of these brines determine whether organisms occur or not. It is odd that these brine organisms are so unknown, as even the ancients were familiar with their various forms and the salt industry today makes use of some of these organisms. It seems that the reason the Dead Sea, for example, is "dead" is not because the salt content is too high, but because the calcium level is too high!

In general, the flora and fauna of the brine can hardly be called poor. The number of living organisms that exist within it is manageable, however, which makes the study of these organisms so interesting. The effect of the environmental factors and their influence on the distribution [of organisms] can be monitored particularly well with this group.

8 Distribution. In order to research the distribution and characteristics of saline organisms, the author has studied a number of brine samples and salt

---

[5]Not a true gas. In alkaline solutions the total carbonate content may be high, but CO_3^{2-} is the major dissolved form present in the carbonate system.

[6]Original German: *"Bei Salzkonzentrationen ueber 10–15% beginnt die Existenzmöglichkeit für Halophyten zu erlöschen. Salzseen mit höheren Gehalt erscheinen im Ganzen vegetationslos (Totes Meer 19.5%), wenn auch einzelne Formen von Bakterien und Algen bei noch stärkeren Konzentrationen beobachtet worden sind (halophile Schwefelbakterien bis 22%, Chlamydomonas bis 25%)".*

water samples from very different origins. The data presented in this dissertation have been obtained through the study of these samples.

Several of the locations, as well as the nature of the samples, are given in the following table; locations familiar to the author from field research are marked with an (x).

A. Marine brine and salt
   1. Leslie Salt Works,[7] San Mateo, California (x)
   2. Redwood City and Elkhorn Salt Works, California (x)
   3. Pt. Lobos, near Carmel, California (x) and Mussel Point, California (x)
   4. The large Lagoon of Venezuela
   5. Coast of Brazil
   6. The West Indies
   7. Cagliari, Sardinia
   8. South of France
   9. Portugal, near Setubal
   10. Hawaii
   11. Tripoli, Salammbô
   12. Djibouti, Africa
   13. Madura, Dutch East Indies
   14. Java, Dutch East Indies
   15. Bombay, British East Indies
B. Aeolic salt
   16. Sambhar Lake, Rajputana, British India
C. Secondary solutions
   17. Epsom Lake, Washington
   18. Dead Sea
   19. Lakes Eyre and Bumbunga, South Australia
   20. Techirghiol Lake, Romania
   21. Salinen of Zevenburgen, Romania [= Zevenbergen, Netherlands?]
D. Desert salts
   22. Pool near Marina, California (x)
   23. Mono Lake, California (x)
   24. Owens Lake, California (x)
   25. Searles Lake, California (x)
   26. Sand Springs, Nevada (x)
   27. Pyramid Lake, Nevada (x)
   28. Walker Lake, Nevada (x)
   29. Soda Lake, Nevada (x)
   30. Great Salt Lake, Utah (x)

Natural habitats could be determined through field research at the locations marked with (x). Salts and brines received from elsewhere were placed into specific solutions to obtain a large amount of data regarding

[7]Now known as Cargill.

the occurrence of certain organisms. The general rules for the establishment of specific environments were given in Chapter V.

It appears that a large number of organisms can be found in brine – so many that it is impossible to provide a complete summary in this report. An abbreviated list of organisms and where they can be found is given below (numbers are from the table above). Those noted in parentheses are locations found by others.

A. Bacteria (salt and strong brines contain very many species, of which only a few are mentioned here)
   1. Purple bacteria (autotrophic, anaerobic), 1, 3, 13, 14, 16, 20, 21, 23, 24, 25, 26, 27, 28, 29
   2. Green sulfur bacteria (autotrophic, anaerobic), 13
   3. Aerobic sulfur bacteria (autotrophic, aerobic), 4, 5, 9, 19, 23
   4. Sulfate reduction (heterotrophic, anaerobic), everywhere
   5. Denitrification (heterotrophic, anaerobic), everywhere
   6. Red, or "stockfish"[8] bacteria (heterotrophic, aerobic), everywhere
B. Protozoa
   7. Amoebae, everywhere, (Cagliari)
   8. Ciliates, 1, 2, 8, 22, 25, 26, (Brittany)
   9. Flagellates, everywhere, (Wieliczka)
C. Worms
   10. Rotifers, 20, 21, 29, (Zevenburgen)
   11. Nematodes, 22
   12. Cestoda, 2, 11, 30, (Salammbô)
D. Insects
   13. Heteroptera (Corixa, etc.), 1, 2, 22
   14. Flies (Ephydra spp.), 1, 2, 22, 24, 25, 26, 27, 28, 29, 30
   15. Coleoptera (Ochthebius, etc.), (Adriatic Sea)
E. Crustacea
   16. Artemia salina, everywhere, (Lymington)
   17. Tigriopus spp. 3, (Adriatic Sea)
F. Fish
   18. Stickleback, 1, 2, (Lorraine)
G. Green algae
   19. Dunaliella spp., everywhere, (South of France)
   20. Asteromonas, 1, 3, 9, (Crimea)
   21. Chlamydomonas, 17, (many locations)
   22. Ctenocladus (Lochmiopsis), 22, 23, 24, (Kulunda Steppe, Siberia)
H. Blue-green algae, everywhere
I. Fungi
   23. Oöspora halophila, most likely cosmopolite, (Utah)

---

[8]Presumably Baas Becking is referring to red halophilic bacteria that are sometimes associated with salted fish.

(a)   (b)

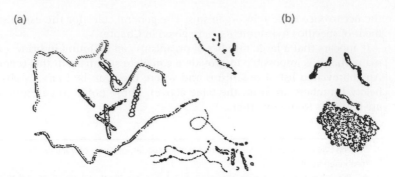

Figure X.6 (a) Colorless sulfur bacteria from seawater taken from Venezuela, Brazil, and Portugal, × 500. (b) Purple spirillum (*Rhodospirillum halophilum* B-B) from Searles Lake, California, × 500. (orig.)

Having reviewed this table, it is clear that many saline organisms have a cosmopolitan distribution. It must be noted that many organisms cannot endure a prolonged stay in dried salt, meaning that the results obtained with older samples are likely not complete. It is remarkable that of the 24 organisms listed here, eight can be called completely cosmopolitan. Most of the restrictions to the occurrence of organisms are based on their specificity; for example those of *Asteromonas* (seawater, brine, temperate climate) and of *Ctenocladus* (*Lochmiopsis*) (alkaline brine, temperate climate, low concentration).

The salt organisms are a good example of the concept "everything is everywhere," which becomes even more apparent after further study of a few representatives.

When light and hydrogen sulfide are available in an alkaline environment, genuine purple bacteria live in anaerobic circumstances. They obtain hydrogen sulfide through sulfate-reducing bacteria. The purple bacteria cause the dark red color of the alkaline desert lakes. In the vicinity of Sand Springs, Nevada, this phenomenon is striking, as a salt marsh of nearly 20 km in length is colored by these organisms. Many salt lakes around the world are called "Red Lake." It is possible that one of the organisms most often responsible for this color, the large red spirillum (*Rhodospirillum halophilum*) is identical to the purple freshwater spirillum. The purple bacteria occur in all brine concentrations (Figure X.6b). Figure X.6a shows representatives of the aerobic sulfur bacteria.

(6) Red bacteria. Heterotrophic and aerobic forms that only contain red pigment (as opposed to the purple bacteria, which also contain green pigment) are often responsible for the deep pink color of the brines in salt ponds, mainly with molar concentrations of NaCl higher than 2. Most likely there are two forms that occur worldwide, namely *Sarcina morrhuae* Kl. (a packet bacteria [four cells in a cubic arrangement]) and *Bacterium halobium* Pett, both of which can cause the red coloring of salted fish (stockfish, herring, etc.). This last form, as well as the cyanobacteria which have vacuoles, possibly keep the aerobic cells afloat in the oxygen-poor water. (Petter, 97) (see Figure X.7).

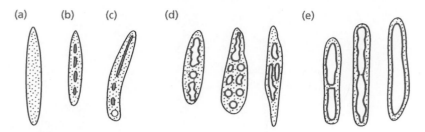

Figure X.7 *Halobacterium salinarium* bacteria of various ages (×1000). Forms with and without gas vacuoles. (From Petter.)

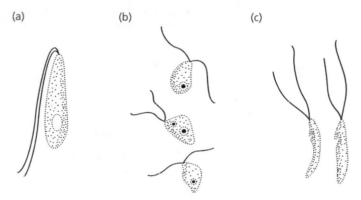

Figure X.8 Colorless brine flagellates. (a) *Amphimonas cuneatus* Namyslovsky (Marina, California). (b) *Amphimonas polymorphus* Namys. (Venezuela). (c) *Pleurostoma parvulum* Namys. (Searles, California).

(4) Sulfate-reducing bacteria (heterotrophic, anaerobic) which form sulfides also occur here. This group of bacteria has a widespread potential environment and is also very adaptable (Elion, 49; Baars, 07). Neither [extreme] temperature nor the salt concentration appear to damage these bacteria, which is why we find a thick layer of black mud at the bottom of all the saline lakes (Searles Lake, 25 meters thick, saturated saline solution).

(7–9) Of the single-celled organisms the amoebae should be mentioned first, as it became apparent that they occur in almost all samples that were not too old. Nonetheless, these organisms, which apparently are not dependent on the concentration of [salt in] their environment, have only been known for some twenty years. Ms. Hamburger first described them from Cagliari.

Colorless flagellates are also widespread (Figures X.8, X.9, X.10). As early as 1916, Namyslovsky (90) described many types from the Wieliszcka salt mines. Geza Entz, Sr. also found a new type in the salinas of Zevenburgen. The author has come across these species and many others in the most diverse regions on earth.

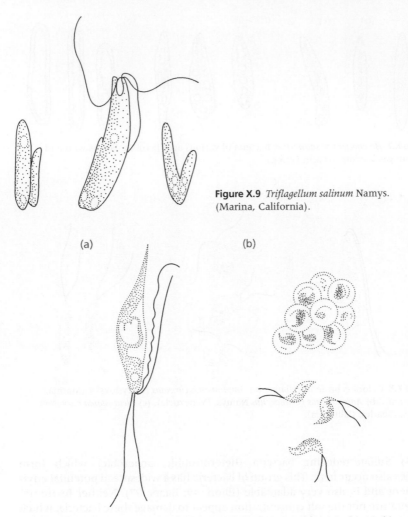

(a)   (b)

Figure X.9 *Triflagellum salinum* Namys.
(Marina, California).

Figure X.10 (a) *Tetramitus salina* (haliphilus) Kirby (Marina, California). (b) *Amphimonas ankyromonadides* Namys. (Setubal, Portugal). All approx. × 1000. (orig.)

(13) Within the Ephydridae, a family of flies, it is mainly species from the genus *Ephydra* (shore fly, or brine fly) which occur globally in saline regions (e.g., *Ephydra millbrae*, Jones). The larva and pupa of this fly can form large brown masses floating on the water, which are sometimes eaten by the Californian Native Americans (72) (Figure X.11).

(15) *Artemia salina* L. In 1756, in the salt works at Lymington (England), Schlosser found a small brine shrimp (up to 12 mm in length) which Linnaeus named *Cancer salinus* that same year. It belongs to the most primitive group of Crustacea, the Phyllopoda. Pallas discovered this creature on his journeys

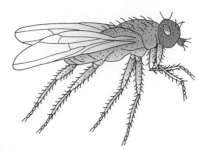

Figure X.11 The Californian brine fly, *Ephydra millbrae*, Jones (Marina, California). × 12. (orig.)

Figure X.12 The brine shrimp, *Artemia salina*; from eggs originating from Marina, California. Male sample × 10. Excrement at the bottom of the image. (orig.)

through Siberia and Russia, d'Arcet came across it in Egypt and in Tunis (where it was mixed with dates and eaten by the locals), while approximately one hundred years ago Anseme Payen described this creature as the cause of the red water in the French salt ponds (Figure X.12). We now know that *Artemia* is cosmopolitan, present in all temperate zones. The eggs can remain dormant for years and are produced in such large quantities that they can be harvested by the pound.

(19) *Dunaliella viridis* Teodoresco and other forms within this genus, green flagellates without a cell wall, are saline organisms *par excellence*, as they occur in salt ponds, desert brines and in both tropical and temperate regions (Figure X.13).

Figure X.13 Green saline flagellate, *Dunaliella viridis* Teodoresco (×4000). Material from Brazil.

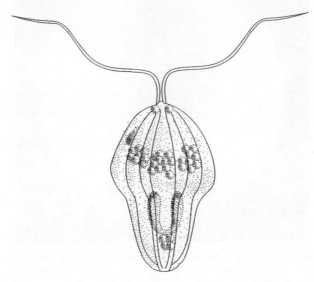

Figure X.14 Green sea-salt flagellate, *Asteromonas gracilis* Artari (×4000). From the Leslie Salt Works, San Mateo, California. (orig.)

It is possible that Pliny had already described saline regions containing *Dunaliella*. Joly proved that the red water in the salts in the south of France was not caused by the brine shrimp *Artemia*, but by its food source, an orange-red form of *Dunaliella*, named *D. salina* by Teodoresco. It is likely that this is a modified form of *D. viridis*. The orange flagellate can occur in huge numbers.

Sven Hedin describes the waters of Lake Kisil-Kul as "seemingly tomato soup, both in consistency and color." This is a very good description of *Dunaliella*!

(20) *Asteromonas gracilis*, yet another bare green flagellate which occurs in a temperate climate (Figure X.14). Artari describes this from the salt lakes

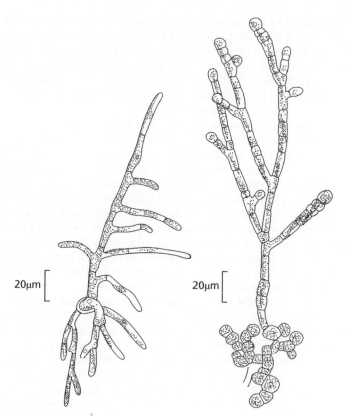

Figure X.15 *Ctenocladus* (*Lochmiopsis*) *sibirica* Woron., in two forms (from J. Ruinen). Left, sibirica; right, Printzii (×500).

adjacent to the Black Sea; the author found this flagellate in very concentrated brine (also magnesium "bittern") from the Leslie Salt Works (San Francisco Bay) and in sea salt from Setubal (Portugal). This form occurs in massive numbers in "rock pools," fissures in the rocks in which large quantities of seagull guano are also captured (14).

(21) *Ctenocladus* (*Lochmiopsis*) *sibirica* Woron. is a filamentous green alga, which, unlike *Asteromonas*, only occurs in alkaline, calcium-poor desert brines in temperate regions. This alga is halotolerant, meaning that its akinetes (survival cells) can endure very high levels of salt concentration, while the germination of these akinetes and sexual reproduction is only possible with concentrations lower than 1 mol NaCl (approx. 6%) (Figure X.15) (128).[9]

Having described these organisms, we can now ask how they influence their environment and how their environment influences them.

---

[9]According to G.#M. Smith 112, this form is identical to *Ctenocladus circinnatus* Borzi.

The interaction of the saline organisms and their environment has been known for a long time (95), although the creatures themselves have only recently been discovered. In the preparation of sea salt, a process perfected by the Chinese long before our time (11), seawater was exposed to sun and wind until it colored red. The red bacteria, which do not develop at lower concentrations, functioned as a living hydrometer, indicating when the point had been reached at which calcium sulfate crystallizes.

The sulfate reduction process has also been long known.

Seventeenth-century English salt producers mention a mud, mixed with sand, black as squid [ink], infecting the whole salt pan when mixed (11). They also provide advice on how, when harvesting the salt (when each millimeter harvested in depth across the whole surface counts), one must exert caution not to harvest the black mud along with the salt, which would color it gray.

In some places in France, Italy, Spain, and Portugal people make use of the cyanobacterium *Microcoleus chthonoplastes*, which covers the bottom of the pans like a layer of linoleum (benthic cyanobacterial mats). The salt on top of this mat is easy to harvest. When preparing a new salt pan, this mat is not forgotten. *Microcoleus*, just like the green algae *Lochmiopsis* [*Ctenocladus*], is halotolerant and can only grow in concentrations below 2 mol NaCl.

Various salt producers still believe the brine shrimp *Artemia* is necessary for preparing good salt. At first glance this seems superstitious. The huge numbers in which the shrimp occur, and the fact that they constantly pass brine through their bodies ("Strudler,"[10] according to Lang), could make the effect analogous to that of the cockle *Cardium*, which fixes the colloidal mud on the Dutch (Wadden Sea) mudflats (105). Van't Hoff (63) described how when seawater is evaporated and the solubility equilibrium of $CaSO_4 \cdot 2H_2O$ has been exceeded, the compound does not precipitate but remains in suspension. Experiments conducted in 1836 by Payen and Audoin, and repeated by the author with the same results, show that fine, stable suspensions (e.g., barium sulfate, calcium carbonate, or calcium sulfate) can be cleared by a relatively small number of brine shrimp (5 per 100 cc liquid) within 24 hours, in which the precipitated colloid is visible as small granules of excrement at the bottom of the vat. They are aptly named "cleaner worms." They help to decalcify the salt. In the pre-reservoir of the saltpans, where the seawater concentrations can be up to 7%, quite a number of sea creatures and seaweed still occur (stickleback, crabs, *Ulva*, etc.).

In sunlight the algae shift the bicarbonate–carbonate equilibrium towards the carbonate end through carbon dioxide assimilation, exceeding the solubility equilibrium of calcium carbonate. Furthermore, the cell walls of many algae passively accumulate calcium carbonate (matrix, *Cladophora*), showing how living systems again exert their influence and speed up the physical–chemical changes to the environment.

---

[10]This is seemingly a passive activity.

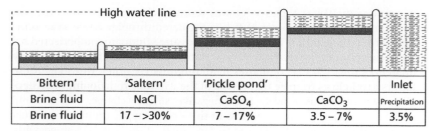

'Bittern'	'Saltern'	'Pickle pond'		Inlet
Brine fluid	NaCl	CaSO$_4$	CaCO$_3$	Precipitation
Brine fluid	17 – >30%	7 – 17%	3.5 – 7%	3.5%

Figure X.16 Schematic overview of a salt pond.

The influence of the organisms on salt production can be examined in Figure X.16. The harvested piles of table salt are exposed to the atmosphere for a long time. This bleaches the colors of the various organisms, while rainwater leaches out the easily soluble magnesium which makes the salt sticky and bitter.

The influence of the environment on the brine organisms brings up so many important questions – mainly physiological – that it is not possible to answer more than a handful within this discourse:

1 The influence of concentration
2 The influence of a shift in concentration
3 The influence of various salts
4 Antagonism of the elements
5 The influence of temperature

1 The influence of concentration. The osmotic environment attracts attention first and foremost. Solutions in which pressures of several hundred atmospheres are found apparently are in a harmonious balance with living systems. This curious fact becomes even more interesting when we examine how the "inner liquid" of such a system is composed. Dr. H. Warren has succeeded in operating on adult brine shrimp with the use of fine glass capillaries to extract unmixed bodily fluids from the creatures. It appeared that the inner liquid never contained a salt concentration higher than 0.9% NaCl, although the outer liquid ranged from 10% to 25%. Such a system cannot easily be explained by the classic theory of osmosis. The osmotic behavior of other salt organisms also remains a mystery. There are many explanations, but as none has been proven, through experiment or otherwise, and the purely physiological reviews fall outside the focus of this discourse, we will not delve any deeper into this material.

Yet some remarks on this topic can be made. Our fellow countryman Schreinemakers and his colleagues have proven the invalidity of the classic theory of osmosis for lifeless systems. It would be tempting to take the rules as deduced by him and transpose them to the salt organisms, were it not for the fact that we can never think of living systems as having physical–chemical *equilibrium*, but only as being systems "in flux." These systems have recently been named "harmonies" by our fellow countryman Straub. It would be more logical to explain the osmotic characteristics of organisms in such a dynamic way. It was Keyes who was recently able to prove that the salt [concentration] in the blood of fish is regulated through active salt excretion. Any consideration of load, swell-pressure, Donnan equilibrium, and the like becomes of secondary importance when the organism actively influences the exchange of matter in an unknown manner. However, these factors can help us to trace the nature of the dynamic functioning. The flagellate *Dunaliella* has a fairly high negative surface charge, which could explain the relative unaffectedness of this organism for anions (negatively charged, thus not accepted by the surface area).

Every salt organism has a specific ratio to the osmotic environment, and it would be presumptuous to regard them all as the same. One must also distinguish between halophilic and halotolerant, not only for specific organisms, but also for their various developmental stages (germination of spores, cell division, formation of gametes, fertilization, etc.). Viewed from this perspective, the solution of the osmotic problem brings up large experimental objections, because of its versatility. For an organism with limited salt tolerance, Ms. J. Ruinen has been able to prove the osmotic behavior of the filamentous algae *Ctenocladus* (107).

2 The influence of a change in concentration, with respect to the varying environment, is of great importance. *Artemia*, in solutions with higher or lower [salt] concentrations, will absorb or excrete water. *Dunaliella* does the same; a percentage [of organisms] may even burst in extremely diluted brine solutions. Red bacteria also swell and shrink (depending on the [salt] concentration). True plasmolysis occurs only in cells which have a vacuole (*Ctenocladus*).

Ms. H. F. M. Petter (97) was able to demonstrate for the red bacteria, and Hof and Frémy (62) for the cyanobacteria, that water extraction through osmotically active materials can be observed through the microscope. Here it can be seen that the red bacteria do not react to NaCl, while a concentrated sugar solution causes a contraction of the cell wall and swelling of the protoplasm. Cyanobacteria [Figure X.17] show signs of actual plasmolysis, also with saline solutions, whereby the protoplasm recedes from the cell wall. In both cases the process is reversible.

From these simple experiments it again becomes clear that each organism can be totally different, and that nature has reached saline resistance through various routes. At this point something should be stated regarding "adaptation." Baars (7) succeeded in habituating the sulfate-reducing bacteria grown at room temperature and in fresh water to high temperatures and a high salt content.[11] Oesterle and Stahl (92) recently conducted a thorough study of the adaptability of the soil bacteria *Bacillus mycoides*. Placed in various external environments, this bacterium displays completely different characteristics. They managed, for instance, through illuminating the cultures, to grow a red form of the organism. They also managed to grow the bacteria directly on 32% salt. This adaptability blurs the line between environments and raises the question of whether the concept of the "potential

---

[11]This would not be normal, and one wonders if Baars started from a pure culture.

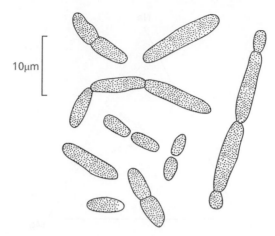

10μm

Figure X.17 Typical cyanobacteria from brine (after Hof and Frémy) (x 1000).

environment" is valid at all. When the difference between the natural environment and the potential environment is taken into account, the potential environment (excluding mutations) should encompass all these possibilities.[12] Only (for example by growing cultures on a certain substrate) a certain area of the potential environment is temporarily difficult to access for a certain organism (halophilization of the red bacteria that will no longer grow with concentrations of NaCl below a certain level). The salt organisms in general are quite adaptable; a large percentage of them can survive sudden changes in [salt] concentration.

3   For brine organisms, the influence of certain salts is also of great importance. In general, NaCl is not toxic and for typical halophiles can be tolerated up to saturated concentration levels. Regarding the other cations, magnesium is also not very toxic. This is valid at least for the larvae (nauplii) of the brine shrimp, for *Dunaliella*, for red bacteria, and for the filamentous algae *Ctenocladus*, although in all of these cases large differences occur. For *Dunaliella* and various colorless saline bacteria it is non-toxic. Red bacteria can only tolerate the high concentrations if sodium is present. For *Ctenocladus* and for the *Artemia* nauplii it is extremely toxic. Calcium on its own is always very toxic, but in combination with other substances it can be more or less detoxified (see later).

Anions have little or no influence on *Dunaliella*. It is remarkable that these algae can be observed for hours swimming in very concentrated solutions of cyanide, rhodanate, or chromate. This can also be observed with adult brine shrimp, which for days can endure the most incredible solutions such as

[12]Most microbes do indeed demonstrate adaptations to a specific range of chemical–physical conditions. This range may be rather broad in some cases, but the idea of a "potential environment" for a specific organism would seem to be valid.

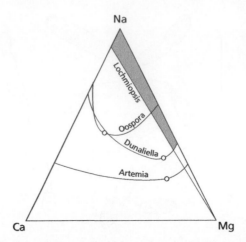

Figure X.18 Potential environment for three chlorides for *Artemia, Dunaliella,* and *Ctenocladus* [*Lochmiopsis*] at 1 molar total concentration. The shaded section shows were all three species can exist. The region for the fungus *Oöspora* is marked for 2 molar concentration.

silver nitrate, potassium dichromate, etc. While for *Dunaliella* (where the naked flagellate is directly in contact with the solution) the effect is direct and physiologically important, the case is different for *Artemia*. As soon as this organism has gone through one or two stages of ecdysis (molting) and has formed a complete intestine it is able to close off the intestine with the sphincter so that only the hard outer shell, consisting of chitin, is in contact with the external liquid. If some barium sulfate is added to the brine one can be assured that it can take days before the transparent intestine is seen to be filled with the white substance. The epithelium that appears to regulate the intake of salts can be found in the central intestine. As long as the shrimp "keeps its mouth shut" it is safe, but note when the liquid [containing the toxin] reaches the cells in the central intestine, and it appears that this organism, at whose "toughness" so many authors have marveled, is just as helpless regarding toxins as any other organism. It is also extremely sensitive to anions, as are red bacteria and, to a lesser degree, *Ctenocladus*.

4 As mentioned above, various substances can "detoxify" each other. This antagonism is of greatest importance for the salt organisms, as it is here that the various substances are subject to the largest changes. With the filamentous algae, *Ctenocladus* for example, 30–80 molecules of sodium are sufficient to detoxify one molecule of potassium, calcium, or magnesium. For hatching of the brine shrimp, for the red bacteria, for the algae *Dunaliella*, and for the fungus *Oöspora* (J. Reuter), the cation limits have been more or less defined as the region in which the individual toxicity of the ions is terminated. Figure X.18 displays in a triangle diagram the various regions in mixtures of Na, Ca, and Mg.

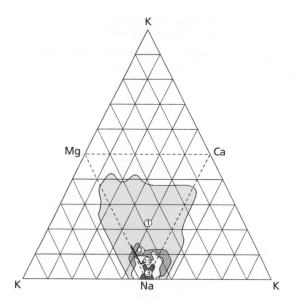

Figure X.19 Antagonism of the chlorides at various concentrations for the hatching of *Artemia*. (After Jacobi and Becking.)

The total salt concentration is 1 molar, and for *Oöspora* (irremissibly halo-philic!) 2 molar. The figure again clearly depicts the great individuality of organisms. The various regions could be seen as "potential environments," whereby *Artemia*, *Dunaliella*, and *Ctenocladus* all occur in the shaded area.

Observations in the field seem to have confirmed this.

The antagonism – which has only been researched for cations, and for only a small number of organisms at that – appears to be highly dependent on the total salt concentration.

In general, the sensitivity for Ca and Mg will increase at higher concentrations of NaCl. This is clearly expressed in Figure X.19, a diagram based on the work of Jacobi on hatching the brine shrimp in various concentrations (70). In this diagram the conventional Na–Ca–Mg diagram is turned 180° and centrally placed, in order to make possible three other triangles showing the combinations Ca–Mg–K, Na–Ca–K, and Na–Mg–K. The lines indicate the borders of the potential environment for combinations of the four chlorides instead of three, for the total concentrations of 1, 2, 2.5, and 3 molar. From the figure it appears that when the concentration is increased the potential environment becomes smaller, resulting in the environment being limited to only table salt (sodium chloride) at a 3 molar concentration. The hook-shaped line in the diagram shows the change in the consistency of seawater with evaporation (compare with Figure X.16). It is the high Mg level of concentrated seawater (3 molar) and not the high salt concentration *per se* which impedes germination/hatching in the natural environment.

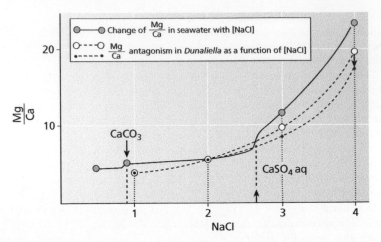

Figure X.20 *Dunaliella*. Antagonism between calcium and magnesium with the evaporation of seawater.

Dunaliella remains alive for longer periods of time in strong brines. This must be attributed to the detoxifying effect of magnesium on the very toxic calcium, whereby this antagonism undergoes an odd change with an increase in salt concentration. When, with the evaporation of seawater in a salt pond (Figure X.20), we consider the molecular ratio of Mg/Ca to be a function of the concentration of the sodium, then this function can be depicted by the solid line in this figure. The line shows two strong Inflections, one when the $CaCO_3$ precipitates, and the other when the $CaSO_4$ precipitates from the solution.

The dashed lines in the figure portray the so called "antagonistic" (detoxifying) values for calcium through magnesium at various sodium concentrations for two species[13] of *Dunaliella*, of which the upper line predicts the values for a species from desert brine and the bottom line the values for a species isolated from seawater brine. It appears that both species behave analogously and are able to exist in seawater brines of various concentrations. Here, it is not that the organism has adapted to its surroundings, but the characteristics of the environment are such that it is especially this organism that can exist here. Also with other external conditions, such as acidity level and temperature, the antagonistic values can change. However, this short discussion suffices here.

5 It is an odd fact that the communities in hot springs and brine show similarities.

At the hot springs near Bingham, the author found the following organisms living at a temperature of 54 °C: purple bacteria, cyanobacteria,

---

[13]In the original, Baas Becking referred to these species as "phyla."

amoebae, colorless flagellates, green flagellates, aerobic sulfur bacteria, and sulfate reduction (black mud). From heat sources of 50 °C and up it is known that the following organisms occur: fly larvae (Brues) and diatoms, all of which are organisms that also occur in brine.

This similarity must be attributed to the fact that the above-mentioned organisms have a very broad potential environment, which makes them, to name it as such, the biological foundation of practically every ecosystem. As the environment slowly becomes more extreme, the creatures with only a limited potential environment will disappear.

Yet research into the thermo-tolerance of salt organisms does have its appeal, especially since the Dutch countryman, the great Hugo de Vries, stated in his dissertation that even with higher plants thermo-tolerance increases when cells are placed in saline solutions.

In general it appears that the salt organisms do not behave abnormally; their tolerance does not exceed 40–50 °C. The optimum temperature is never the same for any two organisms: *Ctenocladus* has a low optimum temperature (18 °C), *Artemia* (at hatching) has an optimum of 25 °C, while the red bacteria develop best at 37 °C.

However, much higher temperatures can be tolerated. A concentrated brine from Searles Lake (California) containing cyanobacteria, green and colorless flagellates, as well as purple bacteria, was heated to 90 °C and observed (on a heating table). It seemed that the organisms not only tolerated the 90 °C heat for almost two hours (after which the experiment was stopped), but remained active and mobile. This experiment was repeated with the same brine solution, diluted and acidified. The results were the same. It turned out that the whole community is thermo-tolerant, or became thermo-tolerant, as neither later experiments with brine from Searles Lake nor experiments with brine of a different origin gave the above-mentioned results; the organisms all died at temperatures below 50 °C! Even though the "why" eludes us, the fact remains that under certain conditions the brine organisms can endure extremely high temperatures.

As with an extremely complicated mathematical equation which only shows its true nature in its boundary conditions, the extreme environment also teaches us to value the external circumstances in an accurate and clear way. Here we find a group of living creatures all together in the same solution which can continuously change in consistency [proportions] and characteristics. Some organisms will disappear and others will appear during these changes. Through the changes in the potential environment – which is fixed – and without taking into account sudden adaptation, the potential environment is traversed point by point by means of the changes in the natural environment. The potentials of living organisms are marked clearly, as while the saturated table-salt solution does no damage to the organism, a small amount of potassium or calcium can end the life of a brine organism. One can see life as a flow connecting an inner world with an outer world. Each specific internal world requires its own flow, which can only keep moving when given certain conditions: its own specific environment.

Appendix

A discussion of the more unusual external conditions would not be complete without mentioning the extremely remarkable communities living in crude oil, recently described by Thorpe (118). In the pools in southern California where this oil is stored there is a fly larva, also belonging to Ephydrinae: *Helaeomyia petrolei* Coq. This larva most likely feeds on the bacteria that occur in the oil, which probably get their energy by metabolizing the oil. It is likely that all the water obtained from the bacteria and larva is derived from their metabolisms (compare Chapter V on the metabolism of the wax moth and the clothes moth).

EDITOR'S NOTES

Baas Becking is clearly at his best in this chapter. We can see this expressed in his descriptions of numerous field studies, outlining his intense curiosity as a naturalist and his wonder at the adaptations of life to extreme environments in the natural world. Ostensibly the subject here is the geobiology of brine, and Baas Becking treats us to many keen observations on the specific adaptations of organisms to this extreme environment. But, more broadly, this chapter is about how organisms adapt to the environment. Brines offer a vehicle by which Baas Becking can explore the "potential environment" of an organism within the context of the environment where it lives. This whole discussion has a very modern feel to it. Indeed, Baas Becking makes a number of very important observations. For example, he notes that while some functional groups of organisms, sulfate reducers for example, are near-universal in distribution, others are not, and as environments become more extreme, the ecological diversity of the environments becomes reduced. He states: "As the environment slowly becomes more extreme, the creatures with only a limited potential environment will disappear."

Baas Becking also makes the very interesting observation that "The optimum temperature is never the same for any two organisms" – highlighting the differences between organisms and explaining how individual species have different "potential environments." In this regard, Baas Becking relates a further observation where a species of sulfate reducer adapted to normal conditions of salt and temperature could re-adapt over time to extreme temperature and salt levels. While organisms generally do display a somewhat plastic response to the environment, indeed defining their "potential environment," the ability of a single species to adapt through such extremes would be exceedingly rare. One wonders if the research described here began with a true pure culture. In any event, this chapter highlights Baas Becking's profound interest in placing organisms within the context of their environment – in essence, his view of geobiology.

CHAPTER XI
Review

As soon as the biologist begins to view the phenomena around him either as given units or as causal aggregates that can be analyzed, his preference will lead him either to the field or to the laboratory. The end goal in both cases is to understand the natural environment around him. This goal can be reached through different means: the field biologist will use analytical methods; the laboratory specialist will construct synthetic forms. But many cases illustrate the fact that synthesis simply "sits" better with one person and analysis with another. Sometimes it seems that field ecologists have established an extensive analytical program that hovers above the facts, while upon closer examination it appears that their goal was to outline parameters to new synthetic forms that they believe they have found in nature.

The term *biocenosis*, or *ecological community*, as it is understood in this book, follows from environmental science. One single factor can dominate, or variability (in time) can enable various limiting factors to emerge as dominant factors. Here, the potential environment determines the presence of certain living organisms. Yet, in ecology and sociology, a biocenosis is often seen more broadly and is regarded intuitively as a complete system of complex organisms that "belong" together. The environmental analysis can be partially or completely missing from this view.

The biological nature which makes it possible to recognize such units [ecological communities] can also be found in the methodology. For many, the concept of "species" is still one of an intuitive unit taken from nature. The insight needed to clarify this concept encompasses an analytical study of the hereditary characteristics [of individual "species"] and of the distribution of species on earth, after which one can again "go into the field," but now with open eyes.

This is the return to the Earth "as it is," as was discussed in the Introduction.

The author finds it incorrect to assume that there is a spiritual gap between those who regard nature as being synthetic (divided into units) and those who view nature analytically (as causal complexes).

The extreme analyst will study living creatures as if they were physical–chemical systems; the extreme synthesist will approach the most ideal description of form: so smooth, so hard, and so rounded, such that the pincers of experiment cannot take hold. Most likely, these extremes exist only in our imaginations; there is no biologist who does not acknowledge the possibilities of experiment *and* of field observations. In this book the analytical side of a synthetic problem has been brought to the fore. Perhaps for this reason the contrasts of the analytical image are sketched more clearly and the relations

Baas Becking's: Geobiology, Or Introduction to Environmental Science, First Edition. Edited by Don E. Canfield.
© 2016 John Wiley & Sons, Ltd. Published 2016 by John Wiley & Sons, Ltd.

more simply than they are in reality. In no case does it seem possible to "explain" an environment, and this will also not be possible in the future. Because when the chemist has pointed out the heterogeneity of water to us, a new environmental factor has already been born, and with progress in physics and chemistry the image we create of the environment will solidify. Two processes are needed to arrive at these synthetic images. The first process is to do as Liebig did, bringing nature "into the home" and viewing the aquarium as a microcosm. This method is followed in Chapters II–VI of this work.

The second process, as described in Chapters VII–X, utilizes the data obtained. The idea behind this method is expressed by M. W. Beyerinck as: "In addition to the infusions of science are those of nature." These infusions, these aqueous extracts from the Earth's crust, have been very incompletely described in this work.

Exploring an organism's potentials for life is an extremely complicated task; the limits of these possibilities, which bound the potential environment, are not known for any organism.

The "synthetic" viewpoint will, often prematurely, declare an organism to be "adapted to its environment," without even having begun to examine the organism's biological possibilities.

For the author, the analytical review of the environment (albeit synthetically inspired) is the only way to come to a synthesis.

EDITOR'S NOTES

This is a beautiful summary for Baas Becking's book and a perfect, and prescient, view of geobiology as a field. Baas Becking so clearly sees that geobiology works as an interplay between experiments and observations; the observations inform the experiments, and the experiments help give meaning to the observations. Always the philosopher, Baas Becking also clearly sees that practicing geobiologists have different natural abilities and tendencies, with their own predisposition to approach their science either as observationalists or as experimentalists. However, Baas Becking also clearly notes that no practicing scientist operates at the extremes of these approaches, so geobiology will always progress through the interplay of approaches. He observes that it is impossible to fully explain a natural environment and an organism's adaptation to it, and that this will always be the case. While making this claim fully 80 years ago, it still rings true today. As Baas Becking points out, the chemical environment is almost endlessly complex in its composition, and this complexity changes in both space and time. Also, organisms interact with the chemical environment, and with other members of the ecological community, in ways that are difficult to fully understand. Indeed, one expression of this is the difficulty of isolating environmentally relevant microbes from nature. In some cases, it seems microbes are so intimately dependent on other ecosystem members that they cannot be isolated independent of these other members. In other cases, it seems that organisms are dependent on the concentrations of specific chemical substances in the environment in ways that are difficult to fully understand and to reproduce in the lab. These problems, however, should be viewed as challenges, and they underscore Baas Becking's famous statement that "*everything is everywhere*: but *the environment selects*." All in all, with these closing comments, Baas Becking gives us a wonderful look into the future of geobiology as a field.

The editor is grateful to Donald Klein and Peter Baas from Colorado State University for pointing out that this all-important chapter was missing in the original published translation. This omission was entirely the fault of the editor.

Appendix

Titles that appeared in this scientific series along with L. G. M. Baas Becking's
Geobiology: Or Introduction to Environmental Science

No. 2:   DE BOUW EN ONTWIKKELING DER STERREN, door Prof. Dr. A. Pannekoek, met illustraties. *f*2.75, gebonden *f*3.75.

No. 4:   ELECTRISCHE ONTLADINGEN IN VERDUNDE GASSEN, door Prof. Dr. Jhr. G. H. Elias (Diligentia cursus). *f*0.90.

No. 7:   DE BEWEGINGSMACHINE VAN HET DIER, door Prof. Dr. H. J. Jordan, met vele illustraties (Diligentia cursus). *f*2.75, gebonden *f*3.75.

No. 8:   DE BOUW DER MOLECULEN, volgens de theorie van Kossel, door Dr. A. E. Van Arkel, met teekeningen en tabel der thans bekende elementen, met vele illustraties. *f*2.75, gebonden *f*3.75.

No. 9:   VIJF EN TWINTIG JAREN MUTATIE-THEORIE, door Prof. Dr. Th. J. Stomps, met uitgebreide literatuurlijst (Diligentia cursus). *f*2.75, gebonden *f*3.75.

No. 10:   MODERNE ZIELKUNDE, door Dr. Th. Van Schelven. *f*2.75, gebonden *f*3.75.

No. 11:   DE FUNDAMENTEN VAN HET SCHAAKSPEL en hun beteekenis voor de praktijk, door Dr. M. Euwe, met 60 diagrammen. *f*2.75, gebonden *f*3.75.

No. 12/13:   HET ORGANISME IN WORDING, door Dr. M. A. Van Herwerden (Lector Universiteit te Utrecht), met zeer vele illustraties. *f*3.90, gebonden *f*4.90.

No. 14:   DE GEOLOGIE VAN NED. INDIE, door Prof. Dr. L. M. R. Rutten (Diligentia cursus), met vele illustraties. *f*2.75, gebonden *f*3.75.

No. 15:   BEKNOPTE VOLKENKUNDE VAN NED.-INDIE, over geboorte, huwelijk, overlijden en godsdienstige gebruiken der inheemse bevolking, door Ph. C. A. J. Quanjer, met illustraties. *f*2.75, gebonden *f*3.75.

No. 16:   DE VRIJMETSELARIJ, OORSPRONG, WEZEN EN DOEL, door A. F. L. Faubel, met illustraties en register (tweede herziene druk). *f*2.75, gebonden *f*3.75.

Baas Becking's: Geobiology, Or *Introduction to Environmental Science*, First Edition. Edited by Don E. Canfield.
© 2016 John Wiley & Sons, Ltd. Published 2016 by John Wiley & Sons, Ltd.

No. 17:	GOLFMECHANIKA, door Prof. Dr. P. Ehrenfest, bewerkt door Dr. H. Casimir, met teekeningen. $f2.75$, gebonden $f3.75$.
No. 18/19:	GEOBIOLOGIE of inleiding tot de milieukunde, door Prof. Dr. L. G. M. Baas Becking, met vele illustraties, literatuurlijst en index. $f3.90$, gebonden $f4.90$.
No. 20:	THEOSOPHIE, door W. A. L. Ros-Vryman. $f2.75$, gebonden $f3.75$ ter perse.
No. 21/22:	EFFECTEN, door Mr. W. M. J. van Butterveld, met afbeelding en register. $f3.90$, gebonden $f4.90$.
No. 23:	KOSMOS, door Dr. W. De Sitter, met illustraties. $f2.75$, gebonden $f3.75$.

References

The numbers marked with an asterisk (*) indicate literature sources that are particularly comprehensive and/or detailed.

1 Abbot, C. G., Fowle, F. E., & Aldrich, L. B. (1913). *Ann. Astroph. Obs. Smiths. Inst.* **3**, 142.
2 Abbot, C. G. (1922). *Ann. Astroph. Obs. Smiths. Inst.* **4**, 203.
3 Armstrong, H. E. (1908). *Proc. Roy. Soc. Sla.* **80**.
4 Arrhenius, Sv. (1908). *Das Werden der Welten*. Leipzig, Akad. Verlag.
5 Atkins, W. R. G., & Harris, G. T. (1924). *Proc. Roy. Dubl. Soc.* **18**, 281.
6 Atkins, W. R. G. (1932). *Journ. Cons, pour l' Explor. Mer* **7**,171.
7 Baars, J. K. (1927). *Over sulfaatreductie door bacteriën*. Diss. Delft.
8 Baas Becking, L. G. M. (1932). *Nature* **130**, 852.
9 Baas Becking, L. G. M., & Tolman, C. (1927). *Econ. Geol.* **22**, 356.
10 Baas Becking, L. G. M. (1927). *Over de algemeenheid van het leven*. Openbare Les, Utrecht.
11 Baas Becking, L. G. M. (1931). *Scientific Monthly.* **32**, 434.
12 Baas Becking, L. G. M & Bakhuyzen, H. v. d. Sande, & Hotelling, H. (1928). *Verh. Kon. Akad. Amst.* **25**, 1.
13 Baas Becking, L. G. M. (1931). *Journ. Gen. Physiol.* **14**, 765.
14 Baas Becking, L. G. M. (1930). *Contr. Marine Biol. Stanford Univ.* **102**.
15 Baas Becking, L. G. M., & Boone, E. (1931). *Journ. Gen. Physiol.* **14**, 753.
16 Baas Becking, L. G. M. (1933). De oorzaak van de zure reactie van het hoogveenwater. Handel. *24e Ned. Nat. & Gen. Congres.* **175**.
17 Barnes, T. C. (1932). *Proc. Nat. Ac. Sc.* **18**, 136.
18 Barnes, T. C., & Lloyd, F. E. (1932). *Proc. Nat. Ac. Sc.* **18**, 422.
19 Baumann, A., & Gully, E. (1910). Die freien Humussäuren des Hochmoores. *Mitt. Königl. Bayr. Moorkulturanstalt.* **4**.
20 Bavendamm, W. (1932). *Arch. f. Mikrobiol.* **3**, 205.
21 Bernard, Cl. (1878). *Leçons sur les phénomènes de la vie.* Paris, Baillière.
22 Beijerinck, M. W., & Jacobsen, H. C. (1908). Influenee des températures absoules de 82 et 20 degrés sur la vitalité des microbes. Prem. Congr. du Froid (Paris).
23 Beijerinck, M. W. (1921). *Verzamelde Geschriften.*'s Gravenhage, Nijhoff.
24 Beijerinck, M. W. (1884). *Nature.* **29**, 3011.
25 Beijerinck, W. (1933). *Sphagnum* en Sphagnetum. *De Levende Natuur* 306 etc.
26 Beijerinck, W. (1926). *Verh. Kon. Akad. Amst.* **25**, 2.
27 *Birge, E. A., & Juday, Ch. (1911). *The inland lakes of Wisconsin.* Wisconsin Biol. Survey.
28 Birge, E. A., & Juday, Ch. (1926). *Proc. Nat. Acad. Sc.* **12**, 515.
29 Black, Ch. S. (1929). *Trans. Wisc. Acad.* **24**, 127.
30 Blackman, F. F. (1905). *Ann. of Botany* **19**, 281.
31 *Braun-Blanquet, J. (1928). *Pflanzensoziologie.* Berlin, Springer.
32 Buch, K. e.a. (1932). *Rapp. & Proc. verb. Cons. pour l' Explor. Mer.* **79**, 1.
33 Buder, J. (1919). *Jahrb. Wiss. Bot.* **58**, 525.
34 *Bunte, H. (1918). *Das Wasser.* Braunschweig, Vieweg.
35 *Cholodny, N. (1925). *Die Eisenbakteriën.* Jena, Fischer.

Baas Becking's: Geobiology, Or Introduction to Environmental Science, First Edition. Edited by Don E. Canfield.
© 2016 John Wiley & Sons, Ltd. Published 2016 by John Wiley & Sons, Ltd.

36 *Clarke, F. W. (1916). *The data of Geochemistry.* Washington.

37 Clarke, F. W., & Washington, H. (1922). *Proc. Nat. Acad. Sc.* **8**, 110.

38 Clarke, J. M. (1921). *Organic dependence and disease.* New Haven. Yale Univ. Press.

39 Cowles, H. (1911). The causes of vegetative cycles.

40 Deines, O. v. (1933). *Die Naturwissenschaften* **21**, 873.

41 Dickson, E. C. c.a. (1925). *Journ. of Infect. Diseases* **36**, 472.

42 Dittmar, W. (1884). *Rept. Challenger Exped.* **1**, 251.

43 Domogalla, B. P., Juday, Ch., & Peterson, W. K. (1925). *Journ. Biol. Chem.* **63**, 269.

44 Domogalla, B. P., & Fred, E. B. (1926). *Journ. Am. Soc. Agronom.* **18**, 897.

45 Donat, A. (1927). *Repert. Spec. Nov.* **46**, 18.

46 Drew, G. H. (1924). *Carnegie Inst. Publ.* **82**.

47 Dumas, J. B., & Boussingault, J. (1844). *Essai de Statistique des Êtres organisés.* Paris, Fortin & Masson.

48 *Eds, Fl. de (1933). *Chronic Fluorine Intoxication.* U. S. Dept. of Agr. Bur. of Chem. (Medicine) **12**, 1.

49 Elion, L. (1924). *Centrallblatt Bakt. II.* **63**, 58.

50 Entz, G. (1884). *Ueber die Infusorien des Golfes von Neapel.* Mitt. Zoöl. Inst. Neapel.

51 Errera, L. (1910). *Recueil d'Oeuvres Philosophiques.* Bruxelles, Lamertin.

52 Escher, B. G. (1920). *De gedaanteveranderingen onzer aarde.* Amsterdam, Wereldbibliotheek.

53 Escher, B. G. (1933). *Wat, hoe, waarom ?* Rede.

54 Florentin, E. (1899). *Etudes sur la Faune des Mares salines de la Lorraine.* Paris.

55 Gistl, R. (1932). *Archiv. f. Mikrobiol.* **3**, 634.

56 Griffioen, K. (1933). *Proc. Kon. Akad. Amst.* **26**, 897.

57 *Harnisch, O. (1929). *Die Biologie der Moore.* Stuttgart, Schweizerbart.

58 *Harvey, H. N. (1928). *Biological chemistry and physics of seawater.* Cambridge (Mass.), Univ. Press.

59 Henderson, L. (1918). *The Fitness of the Environment.* New York, Mc. Millan.

60 Hille Ris Lambers, M. (1925). *De invloed van de temperatuur op de protoplasmastrooming bij Characeeën.* Diss. Utrecht.

61 *Höber, R. (1926). *Physikalische Chemie der Zelle und der Gewebe.* Leipzig, Engelmann.

62 Hof, T., & Frémy, F. (1933). *Rec. Trav. Bot. Néerl.* **30**,140.

63 Hoff, J. H. v. 't, (1912). *Untersuchungen ueber die Bildungsverhältnisse der ozeanischen Salzablagerungen.* Leipzig, Akad. Verlag.

64 Honing, G. (1933). *Onderzoek over de reiniging van zeewater in groote aquaria.* Diss. Amsterdam.

65 Hulburt, E. O. (1928). *Journ. Opt. Soc. Amer.* **17**, 15.

66 Hulburt, E. O. (1932). *Journ. Opt. Soc. Amer.* **22**, 408.

67 *Humphreys, W. J. (1929). *Physics of the air.* New York. Mc Graw-Hill.

68 Imhof, geciteerd uit Simroth, H. (1891). *Die Entstehung der Landtiere.* Leipzig.

69 Irving, L., & Baas Becking, L. G. M. (1924). *Proc. Soc. Exp. Biol.& Med.* **22**, 162.

70 Jacobi, E. F., & Baas Becking, L. G. M. (1933). *Tijdschr. Ned. Dierk. Ver. III*, **3**, 145.

71 Johnston, J. (1916). *Journ. Amer. Chem. Soc.* **38**, 94.

72 Jones, B. J. (1906). *Univ. of Calif. Publ.* **1**, 153.

73 Jong, A. de, (1923). *Microorganisms and low temperature.* Fourth Int. Congr. of Refrigeration. Strasburg.

74 Jordan, H. J. (1918). *Het leven der dieren in het zoete water.* Utrecht, Oosthoek.

75 *Josephus Jitta, N. M. e.a. (1932). *Het kropvraagstuk in Nederland.* 's Gravenhage, Rijksuitgeverij.

76 Kostychev, S. (1931). *Lehrbuch der Pflanzenphy-siologie II.* Berlin, Springer.

77 Liebig, J. v. (1840). *Chemistry and its application to agriculture and physiology.* Report to British Assoc.

78 Lipman, Ch. B. (1921). *Phytopathology* **11**, 301.

79 *Loeb, J. (1908). Physiologische Ionenwirkungen. *Hdb. Biochemie v. Oppenheimer,* **2** (1), 107. Jena, Fischer.

80 Macallum, A. B. (1908). *Trans. Roy. Soc. Canada* **2**, 145.
81 Massink, A. (1923). *Rec. des Trav. chim. Pays-Bas* **92**, 605.
82 *Maucha, R. (1932). *Hydrochemische Methoden in der Limnologie.* Stuttgart, Scheizerbart.
83 Meehan, W. J., & Baas Becking, L. G. M. (1927). *Science* **66**, 42.
84 Miehe, H. (1907). *Die Selbsterhitzung des Heues.* Jena, Fischer.
85 Miche, H. (1930). *Arch. f. Mikrobiol.* **1**, 1.
86 Mitchell, J. P. (1910). A study of the normal constituents of the potable water of the San Francisco Peninsula. *Stanford Publ.* **3**.
87 Molisch, H. (1922). *Populäre biologische Vorträge.* Jean, Fischer.
88 Moravek, V. (1925). *Publ. Sc. Univ. Masaryk.* **53**, 1.
89 Mrazek, A. (1902). *Acta Böhm. Ges. d. Wiss.* **38**, 6.
90 Namyslovski, B. (1913). *Bull. Ac. d. Sc. Grac.* **2**, 88.
91 *Niel, C. B. v. (1931). *Arch. f. Mikrobiol.* **3**, 1.
92 Oesterle, P., & Stahl, C. A. (1929). *Centralblatt. f. Bakt.* **2**, 1—16, 16—25.
93 Pantin, A. A. (1931) nieuwste gegevens in Ellis, W. G. *Nature* **132**, 748.
94 Pearsall,W. H. (1922). *Journ. Ecol.* **9**, 241.
95 Peirce, G. J. (1914). *Carnegie Inst. of Wash. Publ.* **193**, 49.
96 Perfiliev, B. W. (1927). Zur Frage ueber die Rolle der Mikroben in der Erzbildung. (russisch). *Geol. Com. U. S. S. R.* **39**, 795.
97 Petter, H. F. M. (1932). *Over roode en andere bacteriën van gezouten visch.* Diss. Utrecht.
98 Polak, B. (1933). *Verh. Kon. Akad. Amst.* **30**, (3).
99 Polak, B. (1929). *Een onderzoek naar de botanische samenstelling van het hollandsche veen.* Diss. Amsterdam.
100 Pictet, R. (1893). *C. R. Soc. Helv.* **5**.
101 Pringle, J. (1779). *Discourse on the different kinds of air.*
102 Priestley, J. (1775). *Experiments and Observations on different kinds of air.*
103 Pütter, A. (1911). *Arch. f. d. ges. Physiol.* **137**, 595.
104 Rahm, P. G. (1920). *Proc. Kon. Akad. v. Wetensch. Amst.* **23**, 235.
105 Richter, R. (1932). *Bildung künftiger Gesteine in der Gegenwart.* (Lezing, gehouden te Leiden 8 Dec. 1932).
106 Ringer, W. E. (1906). Over de konstantheid in samenstelling van het zeewater. *Jaarb. Rijksinst. Onderz. zee.* **2**, 112.
107 Ruinen, J. (1933). *Ree. Trav. Bot. Néerl.* **30**, 725.
108 *Russell, Sir J. (1923). *The micro-organisms of the soil.* London, Longmans Green.
109 Rutherford, Lord (1933). *Nature* **132**, 955.
110 Schuette, H. A., & Alder, H. (1929). *Trans. Wisc. Ac. Sc.* **24**, 135.
111 Shelford, V. E., & Gail, F. W. (1922). *Publ. Puget Sound Biol. Sta.* **8**, 141.
112 *Smith, G. M. (1933). *Fresh-water Algae of the United States.* New York, Mc-Graw-Hill.
113 Söhngen, N. L. (1906). *Het ontstaan en verdwijnen van waterstof en methaan onder den invloed van het organische leven.* Diss. Delft.
114 Steuer, A. (1910). *Planktonkunde.* Leipzig, Teubner.
115 Stocker, O. (1932). *Salzpflanzen. Hdwb. Natnrw.* **8**, 699.
116 Stockhausen, F. (1907). *Oekologie, oder Anhäufungen nach Beijerinck.* Berlin, Parey.
117 *Thienemann, A. (1933). *Limnologie. Hdwb. Naturw.* **6**, 434.
118 Thorpe, W. H. (1930). *Trans. Entom. Soc. London.* **70**, 331.
119 *Tschermak, A. v. (1924). *Allgemeine Physiologie.* Berlin, Springer.
120 Vernadsky, W. (1924). *La Géochimie.* Paris, Alcan.
121 Vries, H. de (1870). *De invloed der temperatuur op de levensverschijnselen der planten.* 's Gravenhage, Nijhoff.
122 *Waksman, S. (1932). *Principles of soil-microbiology.* Baltimore, Williams & Wilkins.
123 *Walter, H. (1927). *Einführung in die allgemeine Pflanzengeographie Deutschlands.* Jena, Fischer.
124 Walter, H. (1931). *Die Hydratur der Pflanze.* Jena, Fischer.

125 *Weaver, J. E., & Clements, F. E. (1929). *Plant ecology.* New York, Me-Graw-Hill.
126 Whipple, G. C. (1914). *The microscopy of drinking water.* New York, Williams.
127 Winogradsky, S. (1888). *Zur Morphologie und Physiologie der Schwefelbakterien.* Leipzig, Felix.
128 Woronichin, N. N., & Popova, T. G. (1929). *Lochmiopsis, eine neue Algengattung.* (russisch). Bot. Inst. Tomsk.
129 Zirpolo, G. (1932). Les bactéries lumineuses soumises à la température de l'hélium liquide. *Rapp. du Lab. Kamerlingh Onnes.* **35**, 1.

Index

Note: Page numbers in *italics* refer to Figures; those in **bold** to Tables.

Baas Becking's: Geobiology, Or Introduction to Environmental Science, First Edition. Edited by Don E. Canfield.
© 2016 John Wiley & Sons, Ltd. Published 2016 by John Wiley & Sons, Ltd.

Printed and bound by CPI Group (UK) Ltd, Croydon, CR0 4YY

27/10/2024

14580158-0001